C0-CDH-445

A Da Capo Press Reprint Series

FRANKLIN D. ROOSEVELT
AND THE ERA OF THE NEW DEAL
GENERAL EDITOR: FRANK FREIDEL
Harvard University

LANDLORD AND TENANT ON THE COTTON PLANTATION

Division of Research
Work Projects Administration

Research Monographs

Works Progress Administration
Division of Social Research
Research Monograph V

LANDLORD AND TENANT ON THE COTTON PLANTATION

By T. J. Woofter, Jr.

With the Collaboration of

Gordon Blackwell
Harold Hoffsommer
James G. Maddox
Jean M. Massell
B.O. Williams
Waller Wynne, Jr.

DA CAPO PRESS • NEW YORK • 1971

A Da Capo Press Reprint Edition

This Da Capo Press edition of *Landlord and Tenant on the Cotton Plantation* is an unabridged republication of the first edition published in Washington, D.C., in 1936. It is reprinted by permission from a copy of the original edition owned by the Harvard College Library.

Library of Congress Catalog Card Number 77-165691
ISBN 0-306-70337-8

Published by Da Capo Press, Inc.
A Subsidiary of Plenum Publishing Corporation
227 West 17th Street, New York, N.Y. 10011

Manufactured in the United States of America

LANDLORD AND TENANT ON THE COTTON PLANTATION

Research Monographs of
The Division of Social Research
Works Progress Administration

I Six Rural Problem Areas,
Relief—Resources—Rehabilitation

II Comparative Study of Rural Relief
and Non-relief Households

III The Transient Unemployed

IV Urban Workers on Relief

V Landlord and Tenant on the Cotton Plantation

WORKS PROGRESS ADMINISTRATION
DIVISION OF SOCIAL RESEARCH

LANDLORD AND TENANT ON THE COTTON PLANTATION

BY

T. J. WOOFTER, JR.
Coordinator of Rural Research

with the collaboration of

GORDON BLACKWELL
HAROLD HOFFSOMMER
JAMES G. MADDOX
JEAN M. MASSELL
B. O. WILLIAMS
WALLER WYNNE, JR.

RESEARCH MONOGRAPH
V

WASHINGTON

1936

WORKS PROGRESS ADMINISTRATION
HARRY L. HOPKINS, *Administrator*

CORRINGTON GILL
Assistant Administrator

HOWARD B. MYERS, *Director*
Division of Social Research

LETTER OF TRANSMITTAL

WORKS PROGRESS ADMINISTRATION

Washington, D. C., December 1, 1936

Sir:

I have the honor to transmit the findings of a study of landlord-tenant relationships conducted in the seven southeastern cotton States by the Federal Emergency Relief Administration. This report includes social and economic data relating to persons who have been of particular concern to the administrators of programs of rural relief and rehabilitation. The findings are basic not only to an understanding of relief problems but also to the reconstruction of agrarian life in the South, and have bearing on tenancy problems in other areas.

The study was made in the Division of Social Research, under the direction of Howard B. Myers, Director of the Division. The collection and analysis of the data were done under the supervision of T. J. Woofter, Jr., Coordinator of Rural Research, with the assistance of Gordon Blackwell and Waller Wynne, Jr.

The report was prepared by T. J. Woofter, Jr. and edited by Ellen Winston. Special acknowledgment is made of the contributions of the following persons who collaborated in the preparation of certain chapters: Jean M. Massell, chapter V, Credit; James Maddox, chapter VI, Income; B. O. Williams, chapter VIII, Mobility; Harold Hoffsommer, chapter IX, Education; and Gordon Blackwell, chapter X, Relief and Rehabilitation.

Respectfully submitted,

CORRINGTON GILL
Assistant Administrator

Hon. HARRY L. HOPKINS
Works Progress Administrator

CONTENTS

TEXT TABLES

TEXT TABLES CONTINUED

TEXT TABLES CONTINUED

TEXT TABLES CONTINUED

FIGURES

FIGURES CONTINUED

FIGURES CONTINUED

SUPPLEMENTARY TABLES

(Appendix A)

SUPPLEMENTARY TABLES CONTINUED

SUPPLEMENTARY TABLES CONTINUED

SUPPLEMENTARY TABLES CONTINUED

Page

SUPPLEMENTARY TABLES CONTINUED

ILLUSTRATIONS

LANDLORD AND TENANT ON THE COTTON PLANTATION

INTRODUCTION

Presentation of the human elements associated with the land tenure system in the Eastern Cotton Belt is the primary object of the present study. The technical phases of southern agriculture and of farm economics are subordinated in order to focus the discussion on the landlord and the tenant, to describe their relationships, and to analyze the effects of the depression and of the beginnings of recovery on these relationships. To accomplish this the plantation has been made the unit of the study, for the plantation is both an organization for production and the mechanism for the distribution of the product. It constitutes a community within which the tenants and laborers have definite relationships, both with the landlord and among themselves. Table 2 gives a description of these relationships.

The detailed analysis which follows does not present a complete picture of landlord-tenant relations in the Southeast, since the field study of 646 plantations on which it is based was limited to medium-sized and large cotton planting operations[1] in the Eastern Cotton Belt.[2] A tract with five or more resident families, including the landlord, was defined as a plantation;[3] tracts with fewer families were not included in this survey.

The chief difference between a sample embracing purely plantation tenants and one which also includes the tenants on smaller operations is that of efficiency in cotton production. This is measured by the fact that the average yield of cotton in 1934 on all farms in the 7 States was 215 pounds per acre and on enumerated plantations it was 257 pounds per acre.[4] The higher production does not necessarily mean a proportionately higher income for plantation tenants, due to the additional expenditure for fertilizer incurred in more intensive cultivation.

---

[1] For sampling method, see Appendix B, Method and Scope of the Study.

[2] Counties in which 40 percent or more of the gross farm income in 1930 was from cotton farms. Plantations were enumerated in North Carolina, Georgia, Alabama, Mississippi, Arkansas, and Louisiana. In addition plantation data for South Carolina were secured from other studies.

[3] This definition was adapted from the definition used in the U. S. Census special plantation inquiry of 1910. (*See U. S. Census, 1910*, Vol. V, chapter XII, "Plantations in the South.") The present study included tracts with five resident families of any type, *i.e.*, laborer or tenant, whereas the 1910 study included only those tracts with five or more tenant families.

[4] Data for cotton yield on all farms in the seven States calculated from *Agricultural Statistics, 1936*, U. S. Department of Agriculture, Table 99. Data for cotton yield on plantations calculated for 602 plantations operated by wage hands, croppers, or other share tenants.

The method of selection used in this study excluded the majority of the farms in the South, for, contrary to popular belief, large plantations are not now and never have been the mainstay of southern agriculture. In the *ante bellum* South 70 percent of the farmers were non-slaveholders and 50 percent of the slaveholders owned less than five slaves. Today thousands of comparatively small tracts, operated by from one to four families, include the bulk of the acreage and the majority of the farm operators in the Cotton South.

In the Census plantation inquiry of 1910, only 270 counties in the 7 cotton States were considered plantation counties. There were 33,908 plantations in these counties. All farms in these counties contained 653,607 tenants, and the plantations contained 355,186 tenants or 54.3 percent of all tenants in the counties (Appendix Table 1). The other 45.7 percent were scattered on farms with from one to four tenants. There were in addition 240,000 tenants scattered on small farms in these States, outside the 270 plantation counties. The probable number of plantations in these States in 1935 had dropped to about 30,000.

The smaller farming units are obviously incapable of the economies and efficiency of large-scale organizations described in this study and are generally located on poorer cotton land. Also, some of the facts regarding the landlord-tenant relationship brought out in this report might not be applicable to the smaller farms since the relationship is probably less subject to abuse where a small number of tenants permits more frequent, direct, and personal contacts.

The background material in the various chapters of the report includes characteristics of smaller as well as larger farms, however, and in some respects the findings of the field study are applicable to small units. For example, in the character of supervision and method of dividing the crop, the landlord-tenant relationships are generally similar whether the landlord has 1 tenant or 50.

Furthermore, plantation customs and ideology set the pattern for relationships in smaller farm units. This is true because of the dominance of the plantation in southern rural life. Large planters persistently emerge as the political and economic leaders of the cotton areas. Even if there are only four or five large plantations in a county, the ownership of these considerable properties and the prestige of success on a large scale make it easy for the planters to assume prominence in community control if their personalities fit them for leadership. Add to this a sentimental attachment to land as a symbol of aristocracy and the consequent family ties to the land, and the plantation stands out as the basis for a hereditary oligarchy in southern community life.

SUMMARY

Large-scale cash-crop farming continues today in the same areas of the Southeast that had large slave-holdings and large cotton plantations in 1860. Today, however, it is the Negro or white tenant farmer, rather than the Negro slave, who operates most of the plantation land. The increase in tenancy, and especially in white tenancy, has been the most striking trend in southern farm life in the last 25 years. Since 1910 there has been a marked decrease in owner operators, and a great increase in tenant operators, both in proportion to all farm operators and in actual numbers. The major part of this shift occurred before 1930.

A plantation is defined for purposes of this study as a tract farmed by one owner or manager with five or more resident families. These may include the landlord, and laborers, share tenants, or renters. Except in the case of renters the landlord exercises close supervision over operators, and except in the case of wage laborers each family cultivates a separate piece of land. Owners of plantations of the size surveyed do not constitute the majority of southern landholders. Concentrated in some counties, however, they control the majority of the acreage. In those parts of the South where fairly large operative units prevail, the plantation owners, through their control over large acreages of the best land and of large numbers of tenant and laborer families, still dominate the economic, political, and cultural life. Landlord-tenant relationship on the smaller units in such areas are patterned after those on the larger holdings.

Plantation Areas

The cotton plantation areas of today, as in 1860, are regions where the land is adaptable to large-scale cotton production. It is in these areas that a large proportion of Negroes is found in the population. They include the Atlantic Coast Plain cotton-tobacco areas, with medium-sized plantations; the Old Black Belt cotton area, with some remaining large plantations, many of them operated by absentee owners; the Upper Piedmont; the Muscle Shoals Basin; the Interior Plain of Arkansas and Louisiana, with small plantations scattered widely among the smaller farms; and the Mississippi Bluffs and Delta areas, including the bottom land of the tributaries of the Mississippi. In the last named areas the plantation organization has been retained most persistently, and here also the land is particularly well adapted to large-scale cotton production. In these areas, the proportion of Negroes in the population is particularly large.

In true plantation areas there is a high degree of concentration of land ownership, with a consequent high proportion of tenants among the farm operators. Such areas are further characterized by per capita incomes higher than those in other southern agricultural counties but lower than those in other farming sections of the Nation; small proportions of urban and village dwellers; scarcity of non-agricultural industries; large families; poor school facilities, especially for Negroes; and a highly mobile population, with families frequently on the move in search of better conditions. These areas are utterly subject to King Cotton, booming when the King is prosperous and slumping when the King is sick. Aside from feed for livestock and a limited amount of produce for home consumption, practically no other crop is grown.

Labor Conditions and Tenure Classes

As land resources are now used, plantation labor conditions and population trends are largely determined by the pressure of population on these resources. Concentration on one crop—cotton—demands a large labor supply for only part of the year. Landlords prefer large families to meet the labor demands of the peak seasons, thus encouraging a high birth rate. This high rate of population increase in turn perpetuates the plantation system. Natural increase in southern rural areas, especially of the white population, has been more rapid than in other sections. The Negro birth rate is also high, but the high Negro death rate, particularly among infants, results in a lower natural rate of increase for Negroes than for whites. This surplus labor supply has reduced the bargaining power of the individual plantation tenant, making it increasingly difficult for him to free himself from the plantation system and become an independent farmer.

Before the depression, much of the excess labor (approximately a quarter of a million persons each year) migrated from the rural South to areas where the industrial demand was expanding, so that there was little actual increase in the southeastern population living on farms between 1885 and 1930. Since that date, however, the closing of the industrial labor market has caused a piling up of population in plantation areas, at the same time that the Agricultural Adjustment Administration has restricted the demand for labor in cotton production. Serious problems of relief and rehabilitation have resulted.

Wage labor replaced slave labor on plantations immediately after the Civil War, but share-cropping was soon introduced as a method of labor operations. Most of the plantations are now operated largely by share-croppers—virtually laborers who receive half of the crop in return for working the land. Wage labor continues on a few plantations. Others are operated by various types of tenants, some of whom provide work stock and tools and thus receive a larger share of the crop than the share-cropper and some of whom rent the land outright, paying rent in cash or produce. Often all classes of tenants are found on the same plantation. Of the plantations covered in this study, 71 percent were operated by families of mixed tenure, while 16 percent were operated by croppers, 4 percent by wage hands, 3 percent by other share tenants, and 6 percent by renters.

Prior to 1910, when the acreage of improved land was expanding, wage laborers and tenants were sometimes able to improve their status through saving enough to buy work animals and implements, to rent land outright, or even to buy small tracts of land. In 1860, all Negro agricultural workers in the South were laborers. In 1930, only 29 percent were laborers, 58 percent were tenants, and 13 percent were owners. The 58 percent that were tenants included many share-croppers, whose status was essentially that of laborer.

As population pressure increased, whites began to compete with Negroes for places on the land as tenants and laborers. While the vast majority of white agricultural workers were owners in 1860, by 1930 over three-quarters of a million white families in the Southeast had joined the tenant or laborer class. The proportion of white ownership declined steadily with the increase in white tenancy. Whites now make up the majority of tenants in the Old South, as well as in other parts of the country, although nearly all of the plantations in this survey still had Negro tenants, 53 percent operating exclusively with Negroes, 5 percent exclusively with whites, and 42 percent with both. It is evident, therefore, that white tenants are concentrated on the smaller holdings and Negro tenants on the larger.

Ownership

The number and proportion of large holdings in the South have decreased and the number and proportion of small holdings have increased, reflecting the increasing division of land ownership. The disintegration of large tracts was steady from the Civil War to about 1910. At present there is a tendency to hold large tracts together, especially since so much worn-out land has been dropped from cultivation.

A number of proprietors own more than one tract. Large owners further concentrate operations by renting additional land.

Moreover, about 10 percent of the plantation land in the South is in the hands of large banks, insurance companies, and mortgage companies which acquired it through foreclosures in recent years. About 25 percent of all individual holdings have been acquired within the past 5 years.

Negroes have entered the owner group to a limited extent, there being over 200,000 Negro owner operators in 1910. Like white owners, however, their number has declined since 1910. The size of Negro land-holdings is much smaller than that of white holdings.

Many plantation owners are not experienced farmers, having acquired their holdings by inheritance or foreclosure. In such cases, the owner often places the operation and management in other hands. In this study 6 percent of the plantations sampled were found to be absentee owned and 9 percent of the landlords were classified as semi-absentee, since they made infrequent visits to the plantation. Under absentee ownership land abuse is particularly prevalent and operation is especially unstable in times of crisis. Another characteristic of absenteeism is the extent to which landowners engage, at least partially, in other occupations. In this study, 31 percent of all operators devoted more than one-fourth of their time to occupations other than farming.

Plantation Organization and Management

Indicative of the disappearance of very large plantations is the fact that only slightly over a tenth of the plantations sampled operated 800 or more crop acres, and only about the same number housed 30 or more families. The very large plantations were concentrated in the Upper Delta of the Mississippi and its tributaries, and the adjacent Bluffs section. In the Upper Piedmont and Muscle Shoals regions over 90 percent of the plantations had less than 400 crop acres and less than 10 tenant families. In the Black Belt and Coast Plain sections about three-fourths of the plantations had less than 10 families.

The plantation system requires an abundance of skill, energy, and knowledge on the part of the landlord if his operations are to be successful and his tenants are to make a profit. He must be able to plan and assign the crop acreage to the best advantage, handle financial operations, manage labor, animals, and implements, and supervise marketing and subsistence advances. On the very large plantations there is often the additional management of such supplementary enterprises as commissaries, gins, mills, and shops. Usually the owner or landlord is also obligated to aid in the social and community affairs of his tenants. The large plantation owner or manager is assisted in executing these functions by managers, overseers, and gang bosses. On small plantations all of these functions are performed by one man.

One of the landlord's major duties, and one upon which the success of his operation depends, is the expenditure of the plantation's working capital, in the purchase of seeds and fertilizer, in plantation upkeep, and in the apportionment of subsistence advances to the tenants for food and clothing. This practice of subsistence advances, to be repaid by the tenants when the crop is marketed, is one of the chief trouble spots for the landlord. The supervision of these advances determines the living standard of the share tenant.

Cotton Production Trends

The plantation system is bound up with the cash-crop system. Concentration on cotton increased from the Civil War until the boll weevil invasion soon after 1910. Since 1910 there has been a marked shift of the cotton acreage to the States of Texas and Oklahoma, the combined acreage in these States having increased 100 percent from 1910 to 1930. In 1930, half the cotton acreage of the United States was concentrated in those two States. Cotton acreage had increased 40 percent in Mississippi, Arkansas, and Louisiana, while the eastern plantation States, Alabama, Georgia, North Carolina, and South Carolina, had 5 percent less acreage in cotton in 1930 than in 1910. This decrease was largely due to the disorganization caused by the boll weevil as it passed across the South. In some areas of the Southeast, financial distress became serious 25 years ago and the trend for the cotton producer has been downward for a long time, interrupted by only short periods of high prices. The depression since 1930, therefore, merely added to the effects of previous disasters. Weevil damage caused drastic acreage reduction for a few years, and as each State reduced its acreage, Texas and Oklahoma added to theirs. When weevil disorganization had passed and the Eastern Cotton Belt began to attain its former production, States west of the Mississippi continued to expand their acreage. As a result, the supply of cotton far exceeded the demand. Over-production reached a peak when the 1931 crop of 16 million bales was added to a carry-over of 10 million bales, at the same time that domestic and foreign demand was shrinking. As a result, the price fell to 6 cents a pound in 1932, causing heavy losses to all producers. Only the semi-self-sustaining farmer, or the planter with resources or good credit, could continue to operate.

Land Use and the One-crop System

Under the present system of cash-crop farming, plantation land in the South is used more intensively than land in almost any other section of the country. The average plantation family in this study was allotted 25 crop acres; croppers had an average of 20 crop acres; share tenants and renters had about 25 crop acres; and wage hands, about 45 crop acres per family.

Although spoken of as a one-crop system, the cropping arrangement of the Cotton South is really a two-crop system: cotton for cash, and corn for food and feed. Most plantations have as much acreage in corn as in cotton. Four percent of the total expenditure of the plantations studied, however, was for feed which could easily have been grown on the plantations. Up to the inauguration of the cotton reduction program, the plantations of the South tended to be less and less self-supporting, in contrast with the practices of slave plantations which produced a large proportion of their subsistence needs. Depleted fertility of vast tracts of soil and widespread erosion have resulted from this exploiting of land resources in the interest of cotton cropping for cash returns. Consequently, substantial expenditures for fertilizer are necessary in the cultivation of cotton in the Southeast, except in the Delta areas.

Under the crop reduction program of the A.A.A., cotton acreage was reduced, and between 1933 and 1935 probably more crop diversification was undertaken than during any other period of the South's history. The present cropping system, however, makes such a soil conservation practice difficult to introduce, with the result that the South is one of the major erosion areas of the Nation. Ten percent of the land in the United States classified for retirement from arable farming by the National Resources Board in 1934 was cotton land.

In a prolonged cotton crisis even the most efficient planters are no longer able to operate unless they have ample resources or credit. Although exclusive cotton culture results in heavy losses in bad years, the owners of large tracts still concentrate on this crop because no other use of large-scale tracts is so profitable to the landlord in good years. Only the owners and tenants operating small acreages have any advantage in planting their land to foodstuffs. They can use the produce themselves, but the large planters cannot profitably dispose of all the food that their land is capable of producing.

Credit System

Interest is a substantial item in the budget of plantation expenditures of landlords and tenants and is a major obstacle in the way of financial progress for those in either class. Nearly half of the landlords interviewed for this study had long term debts, mostly in the form of mortgages, averaging more than 40 percent of the appraised value of their land, buildings, animals, and machinery. These long term debts were incurred to meet deficits or purchases of machinery, etc. Most of the landlords had availed themselves of government facilities for mortgage loans.

From 1910 to 1928 the amount of mortgage debt almost quadrupled in the seven southeastern cotton States, and the increases in mortgage debt from 1920 to 1928 were proportionately greater in the South than in any other section of the country. A large number of mortgages have been foreclosed in the past 15 years, and the process was accelerated in the early years of the depression.

Slightly more than half (52 percent) of the landlords interviewed had short term debts incurred to meet production expenses for the 1934 crop. The average amount borrowed per plantation was $2,300, which covered about half the requirements for financing annual production. Interest rates on loans to landlords were high, amounting to 10 percent on government loans, 15 percent on bank loans, and 16 percent on merchant accounts. Combined interest on loans and mortgage debts amounted to almost as much as the landlord's net labor income. The banks were the predominant sources of landlord short term credit, only about 22 percent of the short term loans recorded in this study having been supplied by government agencies. All of the government loans went to landlords, since they held the only available security, the crop lien.

Farmers in the seven southeastern cotton States benefited from emergency crop production and feed loans, obtaining 51 percent of all loans granted in the country from 1931 through 1934. They also received 37 percent of the loans granted in 1933 and 1934 under the production credit system authorized by the Farm Credit Act. Both of these types of loans, however, benefited the plantation owner and cash tenant rather than the share-cropper.

The long term debts of tenants are usually contracted with or through the landlord and are either secured by chattel mortgage on livestock or equipment, or simply carried forward on the landlord's books and added to current borrowing as a lien against future production.

The tenant's short term debts for the current season are usually incurred with the landlord, who provides the tenant's share of expenses and his subsistence advances during the crop season, charging them against future production. Sometimes the merchant makes the subsistence advances. The average time for which advances were made, as shown by this survey, was 7 months, and the average advance was $12.80 per family per month.

In a study of 112 croppers in North Carolina in 1928 subsistence advances were found to be mostly in the form of cash, and to constitute more than 63 percent of the cropper's cash farm income. Interest paid on those advances amounted to more than 10 percent of the croppers' cash income.

Tenant rates of interest are even higher than landlord rates. The merchant and landlord charge high interest rates to compensate for losses due to bad debts. However, the spread between landlord rates and tenant rates is greater than the usual percentage of loss on bad debts.

The high rates of interest involved in this system of credit to share-croppers is one of the major factors preventing their rise on the agricultural ladder. Basing the credit system on crop liens discourages tenants from diversifying their crops, and often forces landlords to market crops at disadvantageous times.

Income

Net incomes in the Cotton Belt are low. In 1934 the average net income per plantation was $6,024. With A.A.A. benefits included, 1934 incomes compared favorably with those for 1929.

This survey indicates that in addition to the size of the plantation, plantation income is related to crop acres per plantation, total cotton acres per plantation, the proportion of crop land in cotton, the productivity of the land, and managerial efficiency. On the average, the larger the plantation, the higher the gross and the net income.

The operator's gross cash income on the 645 plantations studied averaged $5,095. The lowest average gross cash income was received by operators in the Muscle Shoals area, and the highest in the Arkansas River area. On 38 percent of the plantations, it was found that the gross cash income of the operator was less than $2,000; it was between $2,000 and $5,000 on 32 percent of the plantations; and more than $5,000 on the remaining 30 percent.

The average net income of the operators was $2,572, about 10 percent of which was in the form of home consumed products. The net income ranged from an average of $1,340 in the Muscle Shoals area to $7,149 in the Arkansas River area.

The landlord's net income in 1934 was sufficient to pay him 6 percent on his invested capital and about $850 for his labor income. In poor years the landlord is likely to lose heavily on his part of the expenses and also on the tenant income paid in advance for subsistence.

The proportion of net plantation income received by share tenants is very small. In 1934, the average net income per family of the wage hands, croppers, share tenants, and renters on plantations in the 11 areas surveyed was only $309, or $73 per capita.

The average net income per family of wage laborers was $180 for the year, varying from $213 in the Arkansas River area to $70 in the Interior Plain area. The average annual net income per capita in this group ranged from $52 to $92. Share-croppers in this survey, who made up more than half the total number of families, averaged $312 per family, or $71 per capita. Their average net income per capita and per family was highest in the Atlantic Coast Plain area and lowest in the Lower Delta. In the latter area, the croppers' average net income amounted to $38 per person, or slightly more than 10 cents per day. Other share tenants had an average net income of $417 per family, or $92 per capita, the highest of any occupational group.

A majority of the 650 cash renters were in the Black Belt and Lower Delta areas. These areas were among the poorest studied, and consequently the average net income for renters is considerably lower than the comparable average for other share tenants. The average for cash renters was $354 per family. The Upper Delta had the highest average net income for cash renters—$561 per family, or $146 per capita.

The landlord's income and the incomes of the various classes of tenants are determined by different factors. The landlord who operates with wage hands assumes the entire risk, and the entire profit or loss accrues to him. The landlord who operates with share-croppers passes some of the risk to the cropper. However, the landlord still furnishes all the fixed capital and, as with wage labor operation, the landlord stands to lose or gain heavily while the cropper's income is low and steady like that of the laborers. In the case of tenants receiving more than half of the crop, and with renters, the landlord's risk is less and his income per acre lower.

Living Conditions

Fuel and house rent are part of the tenant's perquisites but the houses furnished are among the poorest in the Nation. Unpainted four-room frame shacks predominate. Screening is the exception rather than the rule and sanitation is primitive. In a study of farm housing in the Southeast in 1934, it was found that wells furnished the source of water for over 80 percent of both owner and tenant dwellings.

The low income for large families provides only a meagre subsistence. About one-third of the net income is in the form of products raised for home consumption—a few chickens and eggs, home killed pork, syrup, corn meal, cowpeas, and sweet potatoes. These food items are usually available only in the late summer and fall.

During the months when crops are cultivated, the tenant uses another third of his income, at the rate of about $13 per month, for food—mostly flour, lard, and salt pork—and also for kerosene, medicine, and such clothing purchases as cannot be postponed till fall. Another third is spent for clothing and incidentals, usually soon after the fall "settlement." Thus, by winter, resources are exhausted and "slim rations" begin. Clothing, usually purchased once a year, is of the poorest quality. Often the children do not have sufficient warm clothing to go to school.

On 15 percent of the plantations studied the tenants were required to make all purchases at commissaries operated by the landlord. On another 11 percent of the plantations there were commissaries for optional use. A commissary may be a saving feature for the tenant if the advantage of wholesale buying is passed on to him. If not, it is only an added profit-maker for the landlord.

Few of the tenants in this study had gardens and only 55 percent had cows. The effect of poor housing and meager diet was reflected in the health of the families studied. The lack of balance in diet is largely responsible for pellagra and the digestive disorders that are prevalent in the South. Lack of screening makes the control of malaria difficult.

Mobility

Tenants who have not succeeded in locating on good land or with a fair landlord are continually searching for better conditions, many moving from farm to farm each fall. Although they move often, they do not move far. Most of them remain in the county of their birth or locate in adjoining counties.

The rate of farm-to-farm mobility appears to be closely linked with tenure status. The higher the farmer climbs up the "agricultural ladder" the more stable he becomes.

Mobility within the farming occupation is also relatively common as farmers change from one tenure to another. In periods of prosperity, the tendency is to move up the agricultural ladder, while in years of unprofitable operation there is a tendency to shift down the ladder. Until 1910 there was a net movement upward. Since 1910 croppers have made little progress toward share tenancy, and there has been an actual decrease in the number of renters and owners.

A third type of mobility is the shift from the open country to town as the tenant periodically tries his luck at the sawmill, the cottonmill, or odd jobs.

The evidence indicates that Negro tenants are a more stable group with respect to residence than white tenants. This is probably accounted for, to a large extent, by the fact that there are relatively fewer opportunities for Negroes outside of agriculture and that Negro tenants are more easily satisfied than are white tenants.

The average number of years lived on each farm by white plantation families in this study was 4.8 years, and by Negro families, 6.1 years. White share-croppers lived on each farm for an average of 4.4 years, and Negro share-croppers for 5.6 years.

In a study of farmers in South Carolina in 1933, it was found that white tenants move about once every 4 years, and Negro tenants once every 5 or 6 years. White farm owners move about once in 11 years, and Negro owners once in 12 years.

Education

The education of children in farm families of the Southeast has been sadly neglected, chiefly because of the low tax base in plantation areas. Southern States tax themselves for schools as much per dollar of wealth as do other sections, but the wealth is so inadequate that the resulting revenue provides a very small appropriation per child. In addition, as a result of large families and migration of adults of the productive ages, southern rural districts have a much greater number of children to educate, in relation to the number of productive adults and to the value of taxable property, than do other areas. The agricultural system of the Southeast, which encourages the labor of children during the school term, is a further handicap to education. The families most directly affected by the low educational standards of the Southeast are the tenant farmers and share-croppers, and more especially, the Negro share-croppers.

In order to appreciate the significance of school finance in the South, white and Negro education must be appraised separately. Per capita appropriations for Negro teachers' salaries tend to be in inverse ratio to the percentage of Negroes in the total population. In counties with large Negro majorities, where many pupils can be crowded into one-room schools, per capita expenditures for Negro teachers' salaries are lower than in counties where Negroes are more scattered. In areas of large Negro population such as the Black Belt and Delta counties where white pupils are scattered, per capita expenditures for whites are relatively high, whereas in "white" areas per capita expenditures for white teachers are not so great.

Transportation of pupils to consolidated schools has tended to equalize urban-rural opportunities of white children in recent years, but this improvement has not affected Negro schools to an appreciable extent in most parts of the South. During the depression, educational conditions in the South have been markedly retarded. Salaries of teachers, always low, have been drastically cut, and school terms have been shortened, some States closing their rural schools as early as January 1 or February 1. The Negro school term in both urban and rural areas is much shorter than that of white schools. One-teacher schools, which most directly affect cotton plantation families, showed the greatest discrepancy in length of school term.

The educational level of the Southeast is given national significance by the fact that large numbers of persons born in the Southeast make their life contribution in other parts of the United States.

Relief and Rehabilitation

When compared with other parts of the country, the Eastern Cotton States had a relatively high rural relief rate in 1933, the first year of the Federal relief program. Between April 1934, when the rural rehabilitation program was inaugurated, and June 1935, there was little change in the combined rural relief and rehabilitation rate in this area, while the rate was rising generally in other areas. During this period the Eastern Cotton Belt generally had a lower combined monthly rural relief and rehabilitation rate than other areas.

Several factors contributed to this low relief and rehabilitation rate. The chief factor was the rise in the price of cotton following the launching of the A.A.A. crop control program, which brought relative prosperity to those fortunate enough to have retained status in the farm operator group. An extensive shifting of unemployable cases to the care of county administrative units in this area, and development of an experienced case work personnel to weed out undeserving cases, helped to keep down relief rolls. Negroes, no doubt, were under-represented on relief rolls partly because of discrimination against them, and also because Negroes made up the large majority of families on plantations where paternalistic landlord-tenant relationships, persisting longer than on smaller farming units, served to keep plantation families off relief. Many landlords did not favor their tenants going on relief, viewing this form of assistance as a possible demoralizing influence on a hitherto passive labor supply. On the other hand, some large operators did shift the burden of caring for their tenants to the relief administration, which accounted in part for the relatively heavy relief load in 1933.

Displacement of tenants during the early years of the depression was an important factor in the rural relief situation. Relief grants and rehabilitation loans were necessary where the "furnishing" system ceased to operate, but these grants and loans were relatively few and of small size among plantation families, and more frequent among families on smaller farming units. Plantation families with a relief history received aid for an average of only 3½ months during 1934, relief being only a supplementary means of support and the turnover on relief rolls being very rapid for these cases.

The rural rehabilitation program expanded more rapidly in the South than in other regions and supplied a large number of farm laborers and croppers with work stock, thus giving them a higher tenure status, at least temporarily. Studies of the capability of farm families on relief in several Eastern Cotton States indicate that more than one-half of the families with able-bodied members were considered by the county officials to be capable of participating successfully in a rehabilitation program to the extent of attaining ownership of work stock and farming equipment.

Constructive Measures

Constructive efforts to improve the tenant system must take into account certain basic conditions in the South, especially those relating to the quantity and quality of the population, the inter-regional and international aspects of cotton economy, the type of farm organization to be promoted, and the slowness of fundamental social change. Specific programs in the past have been concerned with improvement of State legislation, submarginal land retirement, soil conservation, crop diversification, production control, and credit reform. The operations of these long range programs are hampered by the tenant system to the extent that the landlord-tenant relationship hinges on a money crop agreement. Direct and work relief have alleviated distress in a wholesale manner. The rehabilitation program has been well adapted to readjust farmers to a self-supporting basis. Still more fundamental is the recent proposal to promote the ownership of family-sized farms.

THE AVERAGE COTTON PLANTATION

(Based on rounded averages for the 646 plantations)

The typical cotton plantation operated by 5 or more families in 1934 included a total of 907 acres, of which 385 were in crops, 63 idle, 162 in pasture, 214 in woods, and 83 in waste land (Figure 1). Approximately 86 percent of the 907 acres was owned by the operating landlord and 14 percent was rented from other owners. Of the crop land harvested,[1] 44 percent was planted to cotton. On the typical plantation the wage hand cultivated 45 crop acres, the cropper 20, the other share tenant 26, and the renter 24.

The plantation had a total value of about $28,700[2] of which $21,700 was in land, $3,900 in buildings, $1,900 in animals, and

---

[1] Excluding crop land harvested by renters.

[2] Excluding operator's residence, gins, and commissaries.

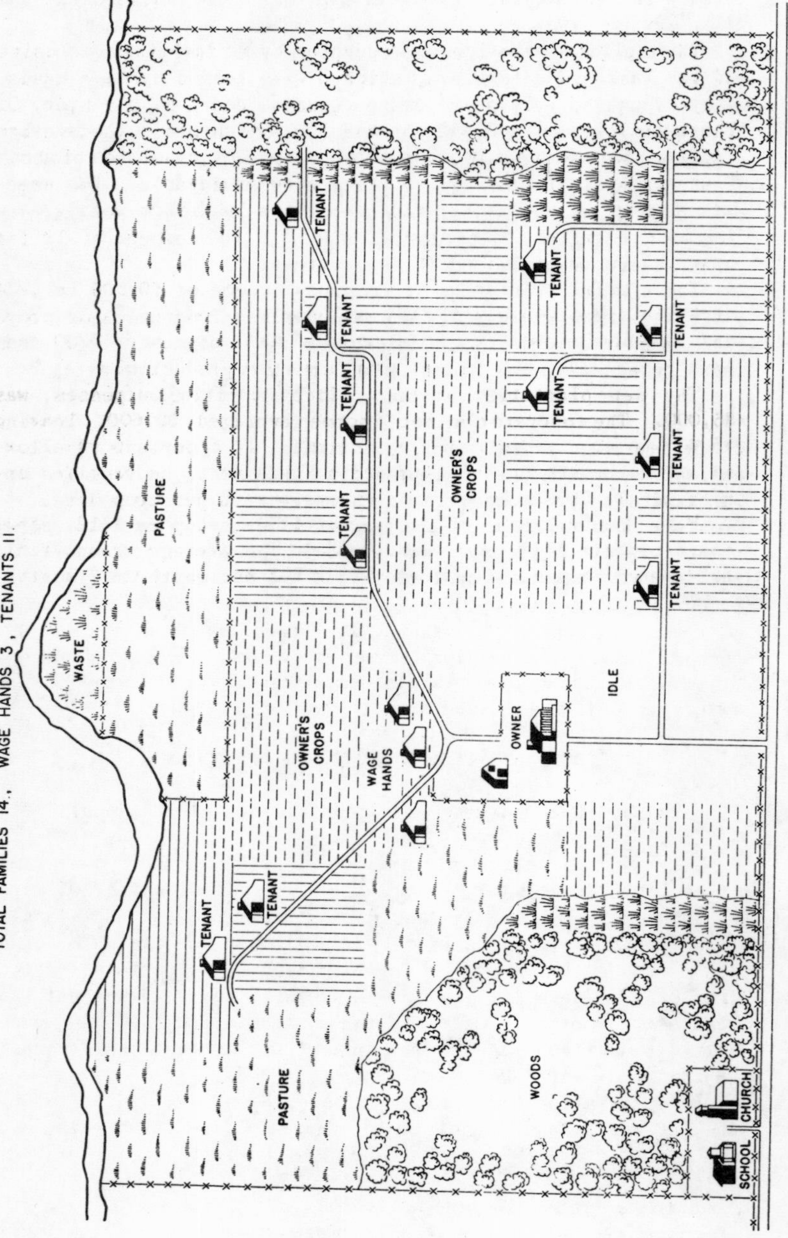

FIG. I THE AVERAGE COTTON PLANTATION (1934)

BASED ON 646 TYPICAL COTTON PLANTATIONS HAVING FIVE OR MORE FAMILIES

TOTAL ACREAGE 907, CROPS 385, WOODS 214, PASTURE 162, WASTE 83, IDLE 63

TOTAL FAMILIES 14, WAGE HANDS 3, TENANTS 11.

$1,200 in implements. The average long term indebtedness was $11,700.

The typical plantation was occupied by 14 families, exclusive of the landlord's family, of which 3 were headed by wage hands, 8 by croppers, 2 by other share tenants, and 1 by a renter. Of these families, 2 were white and 12 were Negro. The average family, the head of which was 41 years of age, consisted of about four persons, of whom two to three were employable. The average number of years of residence on the 1934 farm was 8 years for all families, 7 for wage hands, 7 for croppers, 11 for other share tenants, and 13 for renters.

The typical plantation had a gross income of $9,500 in 1934 of which approximately $7,000 was obtained from sales of crops and livestock products, $900 from A.A.A. payments, $200 from land rented out, and $1,400 from home use production.

The net plantation income, after deducting expenses, was $6,000. The operator's net income averaged $2,600, leaving $3,400 to be divided among the tenants. If 6 percent is allowed as the return on the landlord's investment, he received approximately $850 as his labor income, or $2 per crop acre.

Wage hands had a net income of $180, croppers $312, other share tenants $417, and renters $354. The average tenant family received a subsistence advance of $13 per month for 7 months.

Chapter I

PLANTATION AREAS AND TENANT CLASSES

Ante bellum plantations have persisted as units to a remarkable degree. Some large acreages have been broken up into smaller proprietorships and others have been reduced in size, but in the area characterized by plantations in 1860 large-scale operations persist to a remarkable extent today.

This fact is graphically illustrated in Figures 2 and 3. In Figure 2 large numbers of slaves per owner indicate concentration of plantations in 1860. In Figure 3 a heavy percentage of tenants in the total of farm operators in 1930 indicates similar concentration 70 years later. The coincidence of the areas of concentration is striking. Between these two dates there are two other points of time at which the concentration of large-scale farms can be indexed. In 1900 the Census enumerated the number of rented farms per owner, showing similar concentrations, and in 1910 the special plantation inquiry showed plantations with five or more tenants located for the most part in the same areas.

CHARACTERISTICS OF PLANTATION AREAS

The location of the areas of large-scale ownership and operation is determined by the adaptability of the land to large-scale production, chiefly of cotton (Figure 4). The States of Virginia, Kentucky, and most of Tennessee, as well as the mountainous areas of North Carolina and Georgia, have almost no large-scale tenant operations. The rolling Upper Piedmont section has very few plantations. Likewise few are found in the Muscle Shoals area, the Mississippi Ridge, and the Interior Plain west of the Mississippi, but the level lands of eastern North Carolina, the Lower Piedmont and Upper Coastal Plain of South Carolina, Georgia, and Alabama, and the Delta and Loess Bluff regions of the Mississippi and its tributaries in the States of Mississippi, Louisiana, and Arkansas are regions of heavy plantation concentration.

These areas are characterized by a high percentage of tenants, a high degree of concentration of land ownership, a heavy proportion of Negroes, a very mobile population, per capita farm incomes higher than those in other southern counties but lower than those in other farming sections of the country, small proportions of urban and village dwellers, scarcity of industries,

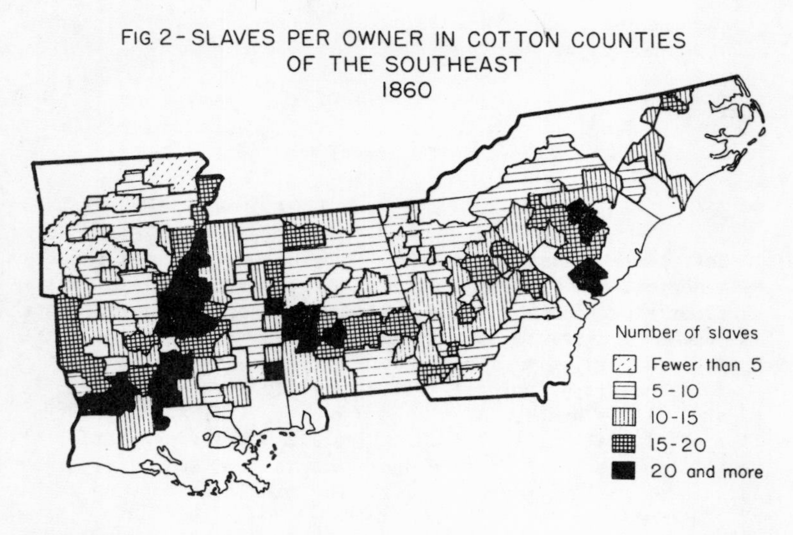

FIG. 2-SLAVES PER OWNER IN COTTON COUNTIES OF THE SOUTHEAST 1860

Source: Eighth Census of the United States: 1860

AF-2021, W.P.A.

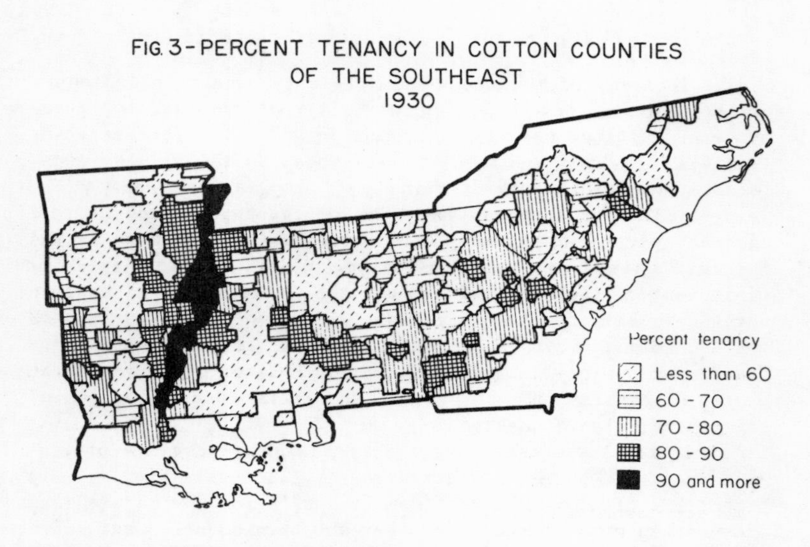

FIG. 3-PERCENT TENANCY IN COTTON COUNTIES OF THE SOUTHEAST 1930

Source: Fifteenth Census of the United States: 1930

AF-2023, W.P.A.

large families, poor school facilities, especially for Negroes, and utter subjection to King Cotton: boom when the King is prosperous and gloom when the King is sick.

Periods of depression in cotton prices are sharp and frequent, fluctuating around the long-time secular trend of general business conditions. Up to the advent of production control there was a close interrelationship between the price of cotton in 1 year and production of cotton in the following year, as well as the obvious relationship of current production to price. The cycle consisted of 1 or 2 years of large production with accumulation of surplus, lowering of price, and subsequent cut in acreage with a rising price, which in turn again stimulated expansion to the point of overproduction.[1]

Plantation Areas Surveyed

The areas of the Southeast which are of particular interest in a cotton[2] plantation study are shown in Figure 5. There is a general cotton-tobacco region in the eastern portion of North and South Carolina. This area has plantations of medium size with only 9.6 percent of the proprietorships in tracts of over 260 acres, according to tax record data obtained for this study in 1935. A similar cotton-tobacco area is found in eastern and southeastern Georgia. These two sections are referred to in the present study as the Atlantic Coast Plain area.

West of the Coast Plain, extending southwest from North Carolina through South Carolina and swinging west through central Alabama into east central Mississippi, is the Black Belt, the oldest area of large plantations. The population of this region includes heavy percentages of tenants and Negroes, and 14.4 percent of the tracts contain more than 260 acres.

West of the Black Belt in North Carolina, South Carolina, and Georgia, is the Upper Piedmont region where the plantations are few, scattered, and small. Only 7.4 percent of the tracts here have more than 260 acres. North of the Black Belt in Alabama and Tennessee is the Muscle Shoals Basin, similar in characteristics to the Upper Piedmont area. Another area of low tenancy ratios and few plantations, resembling the Upper Piedmont, extends north and south through Mississippi and Tennessee and is known as the Ridge or Hill section.[3]

After this ridge is crossed the plantations again become larger and more frequent. The Mississippi Bluffs, an area

[1] Woofter, T. J., Jr., *The Plight of Cigarette Tobacco*, pp. 82-83, Chapel Hill, University of North Carolina Press, 1931. The cycle described in detail for tobacco is also applicable to cotton.

[2] To include all large-scale tenant farming in the Southeast, certain rice and sugar counties in southern Louisiana and certain tobacco counties of the eastern parts of the Carolinas and southeastern Georgia should be added.

[3] Not included in sample because of the small number of plantations in the area.

FIG. 4 - THE COTTON BELT

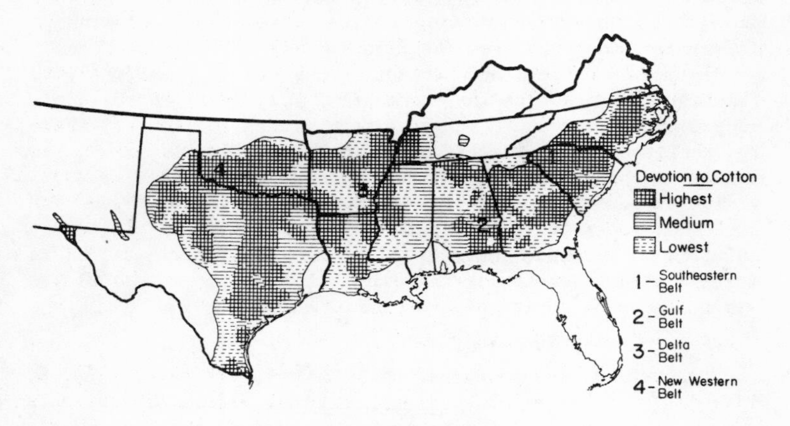

Source: SOUTHERN REGIONAL STUDY, University of North Carolina

AF-2013, W.P.A.

FIG 5 - THE COTTON SOUTHEAST

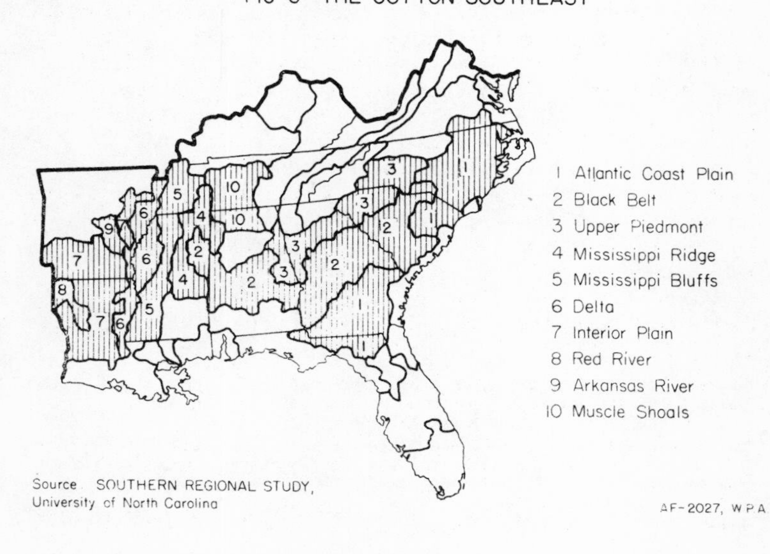

Source SOUTHERN REGIONAL STUDY, University of North Carolina

AF-2027, W.P.A.

several counties wide extending north from the Louisiana line through Mississippi and Tennessee, includes a number of medium-sized plantations with 10.1 percent of the proprietorships containing more than 260 acres.

Along both banks of the Mississippi itself, and extending up its tributaries, the Red, the Yazoo, and the Arkansas, are the Delta or Bottom Lands[4] where large-scale, highly organized plantations persist and are predominant in the rural economy. These last-named areas have the heaviest tenant ratios and heaviest Negro population of any sections in the South. Here large proprietorships make up 17.7 percent of all land tracts, the highest proportion in any area. Another area where plantations are few and very scattered is the Interior Plain of central Louisiana and Arkansas.

Table 1—PLANTATIONS ENUMERATED, BY AREAS

Area	Plantations Enumerated	
	Number	Percent
Total	646	100.0
Atlantic Coast Plain	56	8.7
Upper Piedmont	40	6.2
Black Belt (A)[a]	112	17.3
Black Belt (E)[b]	99	15.3
Upper Delta	133	20.7
Lower Delta	50	7.7
Muscle Shoals	22	3.4
Interior Plain	30	4.6
Mississippi Bluffs	47	7.3
Red River	28	4.3
Arkansas River	29	4.5

[a]Cropper and other share tenant majority.
[b]Renter majority.

Table 1 and Figure 6 show the number of plantations included in the sample from each of these areas.

POPULATION TRENDS AND PLANTATION LABOR

A dominant factor in the southern social and economic system is the pressure of population on resources as the resources are now used. The extensive and almost exclusive use of large areas for cotton culture has limited the expansion of demand for manpower on farms largely to the expansion in cotton acreage, while the increase in consumption of that commodity has by no means kept pace with the expansion of population. In fact, the rate of increase in cotton consumption has hardly exceeded the rate of increase in productivity per man and the expansion of acreage in the Southwest; hence, the increase in number of persons resident

[4]Include Upper Delta, Lower Delta, Red River, and Arkansas River areas.

on cotton farms of the Southeast has not been great since the Civil War in spite of a marked natural[5] population increase in the South. Such increase as has occurred on cotton farms came before 1910. As a result of these factors, the Southeast presents the paradox of too many people for the present system of cash-crop farming, and at the same time a large acreage of idle though fertile land.

Large families have been encouraged by the plantation system because cotton production creates heavy demands for labor in the spring and fall over and above the demands of normal crop operation. Landlords prefer to have this excess labor available on the plantation rather than to import labor for planting and harvesting. They assign a "one horse" or "two horse" crop[6] to a tenant, largely on the basis of the family labor available.

FIG. 6 – PLANTATIONS ENUMERATED

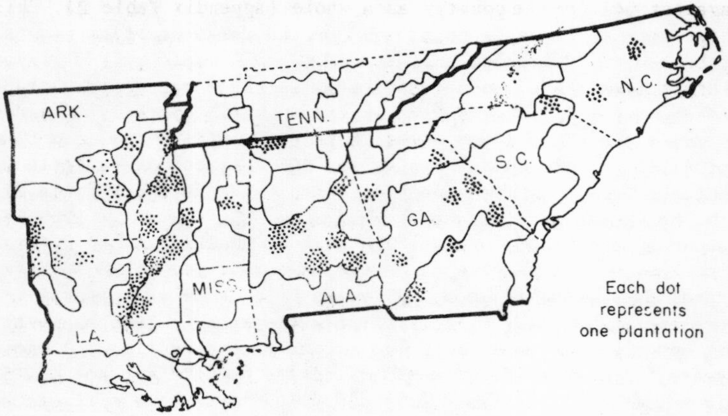

Source: Table I

AF-2025, W.P.A.

Hence, large tenant families have been, at least until recently, an economic asset to both landlord and tenant.[7]

Largely because of the economic advantage of large families, the natural increase of the rural population in the Eastern Cotton States has been more rapid than in any other part of the

---

[5] Natural increase refers to the excess of births over deaths, *i.e.*, the rate at which the population would increase if there were no migration. Actual increase is the result of natural increase *and* migration.

[6] As much acreage as can be cultivated with one or two horses.

[7] There has been some indication that a large family is no longer an economic asset. See chapter X.

country, except in the Appalachian Mountain Area and some parts of the Rocky Mountain Area. The result has been that the needs of the plantation have been plentifully supplied and that there has also been a considerable surplus of labor for other sections. This surplus labor supply has, in turn, reduced the bargaining power of the individual plantation tenant, making it increasingly difficult for him to free himself from the plantation system and become an independent farmer. Thus, it may be said that the plantation system by placing a premium on large families perpetuates a high rate of natural population increase in the South, and that this high rate of increase by producing a surplus labor supply in turn tends to perpetuate the plantation system.

The very rapid natural increase of the southern rural population is shown by the fact that the ratio of children under 5 years of age per 1,000 women 15 to 44 years of age was 591 in the rural farm population of the 7 cotton States in 1930 as against 541 for the country as a whole (Appendix Table 2). This high rate of increase holds true for the rural non-farm population as well. Owing to a higher death rate, especially a higher infant death rate, the Negro natural increase has been less rapid than the white. For native whites the ratio of children under 5 years of age per 1,000 women 15 to 45 in 1930 in the rural farm population of the cotton States was 609 compared with a ratio of 568 for Negroes in the same population group (Appendix Table 2).

The excess of births over deaths in the South in 1930 was about 15 per 1,000, which would mean an annual rate of natural increase of 1.5 percent each year, enough to double the southern rural population in about 45 years if none of the natural increment moved away.[8] Looking back 45 years to 1885, however, it appears that even with the higher rate prevailing in those years,[9] the rural farm dwellers of the South[10] did not double in numbers but increased only slightly. Evidently millions of people emigrated during the generation.

Census figures give some measure of the extent of this migration. They indicate that the rural farm South in the decade 1920 to 1930 exported about a quarter of a million persons each year to cities.[11] Census statistics of birthplace further indicate that 24,100,000 of the native born population of the United States in 1930 were born in the rural Southeast[12] but only

---

[8]Woofter, T. J., Jr., "Southern Population and Social Planning", *Social Forces*, October 1935.

[9]Births and deaths in the South have not been registered for a long enough period to serve as the basis for determining long time trends of population, but the child-woman ratio and such birth and death statistics as are available indicate a slow decline in the rural rate of increase.

[10]South Atlantic and East South Central States, Arkansas, and Louisiana.

[11]Woofter, T. J., Jr., *op. cit.*

[12]Virginia, North Carolina, South Carolina, Georgia, Florida, Kentucky, Tennessee, Alabama, Mississippi, Arkansas, and Louisiana.

17,500,000 of them were living in the area of their birth. Thus, it is evident that over 6,600,000 had moved elsewhere, probably some 3,800,000 leaving the section entirely, and 2,800,000 moving to southern cities.[13] Thus, the southeast rural districts, after supplying their own growth, had exported about a fourth of their natural increase in population, supplying a large proportion of the growth of southern cities, and sending about 3,800,000 to other sections (Figure 7).

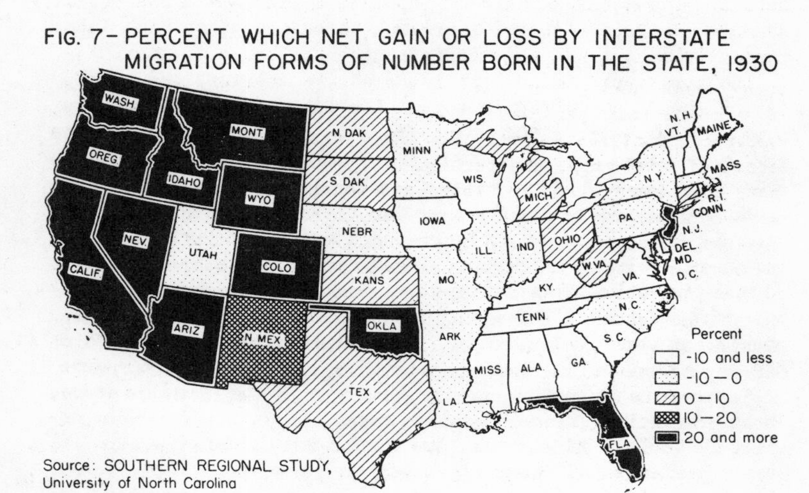

FIG. 7 - PERCENT WHICH NET GAIN OR LOSS BY INTERSTATE MIGRATION FORMS OF NUMBER BORN IN THE STATE, 1930

AF - 2017, W.P.A.

This was the population trend up to 1930. The southern farms were exporting population to the sections where laborers were in demand, first to the West, then to eastern and mid-western industrial cities. Since 1930 the natural increase has continued at approximately the same rate but the urban demand for this excess labor supply has ceased. During the depression years the population piled up in plantation areas and as the Agricultural Adjustment Administration barred the entry of new farmers into agriculture, the problems of relief and rehabilitation in the South were consequently accentuated.

---

[13] 400,000 of the 3,800,000 are compensated for by the movement to southern rural districts from other sections, so that the net loss indicated above is 3,400,000. The small exchange of population between southern cities and rural districts is disregarded in the above calculation.

TENANT CLASSES

The predominant social characteristic of plantation regions is the class-caste system which is built around the landlord-tenant relationship, for tenancy has become not only a method of making a living but also a way of living.

While the plantation proprietorship has continued since pre-Civil War days, merely shrinking somewhat in size, the methods of operation have undergone radical changes. The first of these was, of course, the shift from slave to free labor. The next was the shift from hired labor to half share-cropping, which began very soon after the Civil War.[14] Operation by wage labor continued on those plantations whose owners could afford to finance such operations, but in most instances it was replaced by various forms of tenancy. Share-cropping, in which the farm operator contributes only his labor and receives in return a share of the crop, has persisted, but other forms of tenancy have also emerged. The "third and fourth" arrangement is made with tenants who own their work stock. From them the landlord, instead of receiving half, receives a third of the cotton and a fourth of the corn. These tenants, together with other miscellaneous share tenants, are referred to as "other share tenants" throughout this study. Tenants of a still more independent type rent the land outright, receiving the whole proceeds of their crop minus a fixed rental which may be in cash or produce. These tenants are referred to as "renters" throughout this study.

The principal landlord-tenant arrangements are shown in detail in Table 2 as adapted from Boeger and Goldenweiser.[15] The table presents only the most usual arrangements. There are so many possible modifications and combinations that the relationships are often quite complicated. A tenant may rent from the owner and sub-rent to share-croppers. Mechanized plantations sometimes perform heavy machine jobs for all tenants, charging the expense against their crops or reducing the proportion of the crop allocated to each tenant. In general, however, practically all arrangements are merely modifications of the patterns outlined in Table 2. Thus plantation families are usually readily classifiable into the groups used in this study, *viz.*, wage hands; croppers or half share tenants; other share tenants, who are mostly third and fourth tenants owning work stock; renters; and owners. A sixth group is made up of displaced tenants, former tenants who no longer have a part in the plantation

---

[14]Brooks, R. P., *The Agrarian Revolution in Georgia*, University of Wisconsin, Bulletin 639, 1914, and Woofter. T. J., Jr., *Negro Migration*, W. T. Gray and Company, New York, 1920.

[15]Boeger, E. A. and Goldenweiser, E. A., *A Study of the Tenant Systems of Farming in the Yazoo-Mississippi Delta*, U. S. Department of Agriculture, Bulletin 337, pp. 6-7, 1916.

economy, are not financed by the landlord, but are allowed to live in the plantation houses and often to cultivate some land, mostly for home use production.[16]

Although share-cropping is predominant, all classes of tenants often mingle on the same plantation. Even though the landlord may prefer half share-cropping, he will often take a tenant on terms of third and fourth share or straight rent if he has a

Table 2— SYSTEMS OF TENURE[a]

	Method of Renting		
	Share-cropping (Croppers)	Share Renting (Share Tenants)	Cash Renting (Cash or Standing Tenants)
Landlord furnishes	Land House or cabin Fuel Tools Work stock Seed One-half of fertilizer Feed for work stock	Land House or cabin Fuel One-fourth or one-third of fertilizer	Land House or cabin Fuel
Tenant furnishes	Labor One-half of fertilizer	Labor Work stock Feed for work stock Tools Seed Three-fourths or two-thirds of fertilizer	Labor Work stock Feed for work stock Tools Seed Fertilizer
Landlord receives	One-half of the crop	One-fourth or one-third of the crop	Fixed amount in cash or lint cotton
Tenant receives	One-half of the crop	Three-fourths or two-thirds of the crop	Entire crop less fixed amount

[a]Adapted from Boeger, E. A. and Goldenweiser, E. A., *A Study of the Tenant Systems of Farming in the Yazoo-Mississippi Delta*, U. S. Department of Agriculture, Bulletin 337, 1916, pp. 6–7.

tract vacant, especially when production is expanding. Of the plantations covered in this investigation, 4 percent were operated entirely by wage hands, 16 percent were operated entirely by croppers, 3 percent entirely by other share tenants, 6 percent by renters, and 71 percent were mixed in tenure (Appendix Table 3). On the mixed places, however, croppers predominated.

Both white and Negro tenants were often employed on the same plantation. Of the plantations studied 53 percent were operated entirely by Negro tenants, 5 percent entirely by white tenants, and 42 percent by both white and Negro tenants (Appendix Table 4). In general, the percentage of Negroes in the plantation

[16]This general classification, with the exception of the displaced tenant group, conforms to the usage of the Census before 1920, except that in the earlier years of the Census all share tenants were combined. Recent Census reports separate croppers from other tenants but combine other share tenants with renters who pay rent in kind rather than in cash.

population in each area followed the percentage of Negroes in the rural population.

It must also be remembered that the relations between landlord and tenant are traditionally informal. Detailed agreements are not usually worked out and contracts are practically never written. Such records of advances and repayments as are kept are almost always in the hands of the landlord. This becomes a complicated account when debts from previous years are carried forward and added to current advances. This situation places

Table 3—MALES ENGAGED IN AGRICULTURE[a] IN SEVEN SOUTHEASTERN COTTON STATES,[b] BY COLOR AND TENURE STATUS 1860, 1910, 1930

Color and Tenure Status	Males Engaged in Agriculture (in thousands)					
	1860[c] (Estimated)		1910		1930	
	Number	Percent	Number	Percent	Number	Percent
Total in agriculture	1,132	100.0	2,105	100.0	2,102	100.0
White	325	28.7	1,180	56.0	1,267	60.3
Owners	325		527	25.1	484	23.0
Tenants			418	19.8	581	27.7
Laborers			235	11.1	202	9.6
Negro	807	71.3	925	44.0	835	39.7
Owners			124	5.9	107	5.1
Tenants			477	22.7	486	23.1
Laborers	807		324	15.4	242	11.5

[a]Exclusive of laborers on home farm.
[b]Alabama, Arkansas, Georgia, Louisiana, Mississippi, North Carolina, and South Carolina.
[c]In 1860 there was a very small number of free negro and white tenants.
Source: *United States Census of Agriculture.*

the absolute control of relationships in the hands of the landlord and the fairness of settlements is largely dependent upon his sense of justice. The tenant's only recourse is to move, which of course does not adjust his past transactions but merely enables him to seek more satisfactory conditions.

Thus the prosperity of landlord and tenant are interwoven and mutually dependent upon three principal factors: (1) the productivity of the land, (2) the efficiency and energy of the landlord, and (3) the ability and energy of the tenant. There is evidence that these factors also interact on each other. In their efforts to farm more efficiently the most able landlords tend to get the most productive land. In their wanderings to better their condition the most able tenants eventually gravitate to the fairest landlords. Under these conditions there tends to be a multiplication of supermarginality with the best land, the most capable management, and the most efficient tenants at the top, and a multiplication of submarginality with the poorest land, the poorest managers, and the least able tenants at the bottom.

One effect of the landlord-tenant system as developed in the South was to furnish an avenue through which the landless Negroes and whites could, though with great difficulty, advance to a status higher than that of hired laborer and sometimes even to ownership of the soil. Dr. W. E. B. DuBois in *The Negro Landholder of Georgia* summed up the result so far as the Negro was concerned as follows:

> No such curious and reckless experiment in emancipation has been made in modern times. Certainly it would not have been unnatural to suspect that under the circumstances the Negroes would become a mass of poverty-stricken vagabonds and criminals for many generations; and yet this has been far from the case.[17]

> A thrifty Negro in the hands of well-disposed landowners and honest merchants early became an independent landowner. A shiftless, ignorant Negro in the hands of unscrupulous landlords or Shylocks became something worse than a slave. The masses of Negroes between these two extremes fared as chance and the weather let them.[18]

In 1860 the situation was simple. Practically all land in the Southeast was cultivated by planters with slave labor or by small white owners with their own family labor supplemented occasionally by some hired labor. Hence, at that time practically all whites engaged in agriculture were owners and almost all Negroes were laborers. In the 7 cotton States[19] represented in this study the total males engaged in agriculture[20] increased from about 1,100,000 in 1860 to 2,100,000 in 1930, or 91 percent. This was for the most part a white increase since Negroes engaged in farming increased only about 28,000 or 3 percent, as against a white increment of 940,000 or nearly 300 percent (Table 3).

In addition to the fact that the number of white owners increased about 50 percent, two entirely new classes came into southern agriculture—the white tenants (including croppers and renters) and white hired laborers. Together these numbered, in 1930 in the 7 States, 783,000 white workers who were competing with Negroes for a place on the land. In fact, the most striking trend in the past 30 years has been the increase

---

[17] U. S. Department of Labor, Bulletin 35, 1901, p. 648.

[18] Idem, p. 668.

[19] North Carolina, South Carolina, Georgia, Alabama, Mississippi, Louisiana, and Arkansas.

[20] Excluding members of families working on home farms.

of white tenancy in the South (Figure 8 and Table 3). Tenant conditions can no longer be shrugged aside as features of the race problem, as white and Negro tenants are in most respects equally disadvantaged.

While Negroes were losing their proportionate representation in the total agricultural picture, they markedly improved their status in relation to the land. Though their status upon emancipation was purely that of laborer, by 1930 only 29 percent of the Negroes in agriculture were laborers, 58 percent being

FIG. 8

MALES ENGAGED IN AGRICULTURE

IN SEVEN SOUTHEASTERN COTTON STATES

OWNERS TENANTS LABORERS

1860 VERY SMALL NUMBER OF FREE NEGRO AND WHITE TENANTS

1910

1930

EACH FIGURE REPRESENTS 100,000 PERSONS AF-1562, W.P.A.

SOURCE: UNITED STATES CENSUS OF AGRICULTURE

tenants, and 13 percent owners (Table 3). Among the whites, on the other hand, the proportion of ownership declined steadily with the rapid rise of white tenancy. Thus, the present Negro tenants and owners are children and grandchildren of laborers while the white tenants and laborers are children and grandchildren of landowners. For the former, tenancy is a step in advance of the previous generation, for the latter a step backward.

The most alarming feature of the tenancy trend is the increase during the last 25 years in number and percent of tenants.[21] So long as both the number of tenants and the number

[21] The depression trend from 1930 to 1935 is analyzed separately in chapter X. This analysis shows that the major part of the increase in tenancy occurred from 1910 to 1930 and that there was practically no change in the cotton counties from 1930 to 1935.

of owner operators increased it was fair to assume that some tenants were passing into ownership and that their ranks were recruited from former laborers. This was the case up to 1910 when the acreage of improved land was expanding. Since 1910, however, disasters have been more frequent in the Eastern Cotton Belt and not only has it been more difficult for tenants to accumulate property but they have also felt an increased unwillingness to undertake the financial risks of owner operation. In the 20 years preceding 1930, with practically no change in the improved acreage, the number of owners decreased by 60,000 or 9 percent, while the number of tenants increased 172,000 or nearly 20 percent (Table 3). At the same time the number of laborers decreased 21 percent. It is therefore apparent that there was some shift from ownership to tenancy as well as some continued recruiting of tenants from the ranks of laborers.

When cotton was prosperous laborers and tenants shifted up the agricultural ladder. Croppers purchased work animals and became third and fourth tenants or renters. A few who already owned animals and implements made first payments on land. Likewise, in times when the demand for labor on cotton farms was strong, the tenant was in a better bargaining position. When prices broke or yields were poor, the demand lessened and tenants were placed in a poor bargaining position even if they did not lose their work animals or the equity in land purchased.

Since bad years outnumbered good in the 25 years ending in 1935 the net shift was down the ladder, with losses in ownership and independent renting and large gains in the helpless sharecropper class, fixing the institution of tenancy more firmly in the southern agricultural organization. This arrest in the expansion of the family-sized farm is one of the most fundamental changes in southern rural life, since tenancy not only determines the way in which the soil will be cultivated and the product divided, but also, as subsequent chapters will show, it profoundly influences the personal characteristics of the tenant, his housing and diet, his social contacts, and his institutional advantages.

Chapter II

OWNERSHIP

Land use in the Old Cotton South is conditioned by the fact that there has been little net increase in farm land. Much of the best cotton land was taken up by 1840, and since then it has been almost continuously cultivated in cotton and corn. The rate at which this land has worn out and been abandoned has about balanced the rate at which new land has been drained and cleared.

TREND OF LAND IN FARMS

The disorganization resulting from the Civil War caused a sharp drop in land use up to 1870. From 1870 to 1910 cotton production was expanding slowly and normally to meet the increasing demand. Some forest areas along the Atlantic and Gulf coasts and in the Delta regions were cleared and brought under cultivation. Many plantations which were not operated during the disorganized period after the war came back into cultivation, with the result that land in farms increased.

From 1910 to 1923, however, the seven eastern cotton States were progressively laid waste by the boll weevil. Although each State tended to recover from weevil ravages within a few years after the maximum damage, few of the areas in the Atlantic States recaptured their 1910 acreage, owing to the expansion of cotton culture in Arkansas and Mississippi and the rapid expansion in the Southwest. In these seven eastern cotton States, from 1920 to 1930, there was a loss of 19 million acres, or something over 14 percent of the land in farms. This was accompanied by a slight decrease in improved acreage. That is to say, in addition to idle land within farms there are in the southeastern States millions of acres in abandoned farms.

As a result, there was actually less land within the boundaries of farms in the Southeast in 1930 than in 1860 although the improved or potential crop acreage had increased (Figure 9). The depression-A.A.A. period practically crystallized the situation, although there was a slight increase both in total farm land and in crop land from 1930 to 1935. The increase in total farm land, however, was largely accounted for by an increase in woodland and woodland pasture. Shifts in the size of agricultural tracts owned have been caused by subdivision or recombination of tracts previously used rather than occupation of additional lands, except in special areas such as the Atlantic Coast Plain where pine lands have been cleared.

Various systems of operation of cotton lands have brought about an extremely complex set of relationships between the ownership of the soil, the supervision of the agricultural operations, and the actual performance of the labor. These three functions are combined in a number of ways. The small owner operator performs all three. On the other hand, the resident owner who operates with croppers owns the land and supervises the operation while the croppers perform the labor. The cash

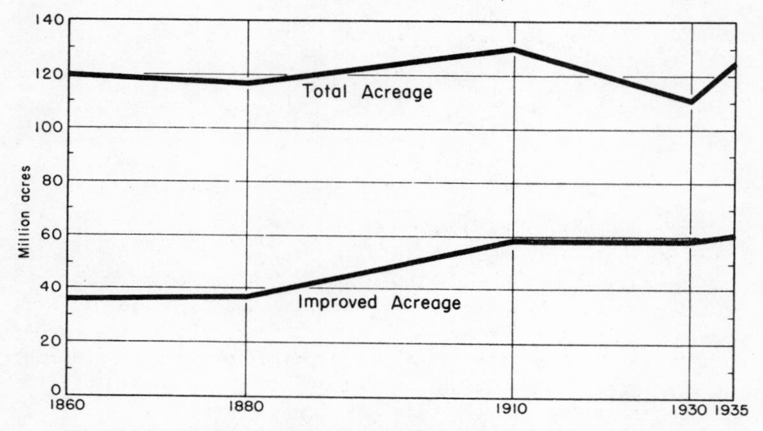

FIG. 9 - TOTAL AND IMPROVED ACREAGE IN FARMS IN SEVEN SOUTHEASTERN COTTON STATES 1860 - 1935

Source: United States Census of Agriculture AF-1485, W.P.A.

renter does not own the land but may supervise the labor of others if the operations are large. The large landholder, if he is able to devote some time to supervision, has a chance of making not only interest on his investment but also profit on financing or managing the operation of each tenant on his land. Hence, there is a tendency to hold large tracts together. The opportunity for the increase in family sized farms comes only through the disintegration of these large tracts or the clearing and draining of new lands.

Some of the categories of relationship of ownership to operation are classified by the Census in its tabulations of laborers, tenants, and owner operators. However, these categories are based on the operation and not on the ownership of the land. A man may own 2,000 acres in a single tract but unless he operates it himself or with hired labor he is not counted by the Census as an owner operator. If he has 50 tenants on his land, working under his supervision, the tract is recorded by the Census as 50 separate tenant farms and no account is taken of the common ownership. Thus Census figures on size of farms do not reflect the size of proprietorships.

The tax books, are, therefore, the principal source of information as to the number and size of tracts owned, regardless of how the tracts are operated. For this study material was obtained from the tax books of 38 typical plantation counties as indicating the trend of plantation ownership.[1] Tracts of 260 acres or more were considered as plantation size, since 260 acres was about the lowest limit of the sample plantations studied. A decrease in the number of tracts above this size on the tax books would, therefore, index the subdivision of plantations while the increase in the number of large tracts would indicate a reconcentration.

In 1934 only 12 percent of the proprietorships in the 38 cotton counties were plantation size (260 acres and over), 53 percent were between 50 and 260 acres, and 35 percent were below 50 acres, most of the last group being one-family farms (Appendix Table 5).

TREND IN OWNERSHIP

Size of Holding

From the Civil War to about 1910, when cotton production was expanding normally and tenants had the opportunity to make enough money to pay for land and some new acreage was being brought under the plow, the disintegration of large proprietorships into smaller ones was steady. Table 4 shows the results of this process in 20 Georgia plantation counties. Tracts of 260 or more acres constituted 38 percent of all tracts in 1873 and 16 percent in 1934. Although most of this change in proportion took place between 1873 and 1902, there was a steady decline in the average size of agricultural proprietorships from 343 acres in 1873 to 185 acres in 1934. Again most of this decline occured soon after the Civil War.

The change in actual number of large tracts has not been so rapid, however, since the division of one large 2,000-acre plantation into 20 100-acre tracts would reduce the percentage of large tracts materially by the subtraction of only one such

---

[1]Advantage was also taken of the fact that E. M. Banks in his *Economics of Land Tenure in Georgia*, Columbia University Press, 1905, had tabulated the size of holdings in 1873, 1883, 1893, and 1902 for 31 counties in Georgia, 20 of which were in the cotton plantation area.

tract. For this reason the increase in the number of small tracts is much more rapid than the decrease in the number of large tracts. The actual number of tracts of 260 acres and over in the Georgia counties sampled decreased only from 4,099 in

Table 4—LAND PROPRIETORSHIPS, BY SIZE, IN 20 GEORGIA PLANTATION COUNTIES,[a] 1873, 1902, 1922, AND 1934

Size of Proprietorship	1873		1902		1922		1934	
	Number	Percent	Number	Percent	Number	Percent	Number	Percent
Total	10,897	100.0	17,010	100.0	20,942	100.0	22,397	100.0
Less than 10 acres	114	1.0	390	2.3	663	3.1	1,432	6.4
10 to 20 acres	79	0.7	386	2.3	636	3.0	959	4.3
20 to 50 acres	716	6.6	1,838	10.8	2,929	14.0	3,337	14.9
50 to 100 acres	1,377	12.6	3,485	20.5	5,150	24.6	5,330	23.8
100 to 175 acres	2,597	23.8	4,290	25.2	5,083	24.3	5,094	22.7
175 to 250 acres	1,915	17.6	2,619	15.4	2,653	12.7	2,599	11.6
250 to 500 acres	2,068	19.0	2,301	13.5	2,254	10.8	2,144	9.6
500 to 1,000 acres	1,315	12.1	1,175	6.9	1,095	5.2	999	4.5
1,000 acres and over	716	6.6	526	3.1	479	2.3	503	2.2
Average acreage per holding	343		243		194		185	

[a]Banks, Butts, Clay, Coweta, Dougherty, Forsyth, Greene, Jasper, Johnson, Lincoln, Madison, Newton, Paulding, Polk, Putnam, Sumter, Talbot, Telfair, Troup, and Wilkes.

Source: Tax digests in the respective counties. Proprietorships 1873 and 1902 compiled by E. M. Banks, *Economics of Land Tenure in Georgia*, 1905; 1922-1934 compiled by the staff of this study.

1873 to 3,646 in 1934, while the increase in the smaller tracts under 50 acres was rapid. There was little change in the number of large tracts from 1873 to 1902 owing to the subdivision of tracts of over 1,000 acres into farms of from 260 to 500 acres.

Table 5—ACREAGE IN PROPRIETORSHIPS OF SPECIFIED SIZE IN 20 GEORGIA PLANTATION COUNTIES,[a] 1873-1934

Size of Proprietorship	Total Acres			
	1873		1934	
	Number	Percent	Number	Percent
Total	3,735,002	100.0	4,146,579	100.0
Less than 10 acres	570	*	7,160	0.2
10 to 20 acres	1,200	*	14,385	0.3
20 to 50 acres	25,060	0.7	133,480	3.2
50 to 100 acres	103,275	2.8	399,750	9.7
100 to 175 acres	361,780	9.7	713,160	17.2
175 to 260 acres	421,300	11.2	569,800	13.8
260 to 500 acres	744,480	19.9	771,840	18.6
500 to 1,000 acres	920,500	24.7	699,300	16.9
1,000 acres and over	1,156,837	31.0	837,704	20.1

*Less than 0.05 percent.

[a]Banks, Butts, Clay, Coweta, Dougherty, Forsyth, Greene, Jasper, Johnson, Lincoln, Madison, Newton, Paulding, Polk, Putnam, Sumter, Talbot, Telfair, Troup, and Wilkes.

Source: Estimated on basis of Table 4.

During the years 1922 to 1934, with a total increase in proprietorships of 1,455, there was a decrease of only 182 large proprietorships, causing a slight reduction in the percentage of tracts over 260 acres in size. In 1873, 37.7 percent of the proprietors held 260 acres or more. In 1934, only 16.3 percent held such large tracts. In 1873, 75.6 percent of all acreage was in tracts of 260 acres and over; in 1934, only 55.7 percent was so concentrated (Table 5).

The averages for the State of Georgia combine Piedmont counties, Coast Plain counties, and Black Belt counties, and therefore mask subregional differences. When the varying distribution of acreage in three States (North Carolina, Georgia, and Mississippi) in recent years is examined (Appendix Table 6), it appears that there was an actual increase from 1922 to 1934 in the number of large tracts in the Atlantic Coast region, where new lands were converted into agricultural use after the removal of large bodies of timber. However, the simultaneous increase in small tracts reduced the percentage of large tracts slightly. There was considerable disintegration of large tracts in the Mississippi Bluffs section. The land trend was almost static in the Piedmont and Delta areas. While there was little reduction in the number of large proprietorships in the Black Belt, there was a rapid increase in small proprietorships, indicating that a number of small farms were carved off very large tracts without reducing the parent tracts below 260 acres. The total effect in the 39 counties sampled in these 5 regions was a decrease of only about 200 out of 8,400 tracts of 260 acres or more.

Multiple Ownership

An additional index to the concentration of ownership is found in the number of landlords that hold non-contiguous tracts. Such multiple ownership is not included in the foregoing figures on size of proprietorships, as each tract is usually carried separately on the tax books. In the present study 39 percent of the landlords reported owning other farms with an average of 2.9 other farms per multiple owner (Appendix Table 7).

It is not always possible to determine whether the holdings of a single owner are returned in one total on the tax records or whether separate tracts are listed separately. For this reason the size of proprietorship statistics and the multiple ownership statistics are merely supplementary measures of land concentration. It is also impossible to obtain for past years the proportion of plantation operators owning more than one tract. For this reason the questions as to reconcentration or continued disintegration of landholdings cannot be accurately answered. The evidence indicates, however, that from 1922 to 1934 there was only a slight decrease in number of proprietorships of 260 acres or more listed for taxation. But in 1934 the ownership of more than one non-contiguous tract was a common practice among the large operators. With this group of large tenant operated holdings, plantation farming partakes of the character of big business.

In addition to concentration of ownership, many owners of large acreages also operate additional rented land. This practice is indicated in the present study by the following distribution of plantation acreage: owned, 86 percent; additional rented, 14 percent.

Another type of concentration has occurred during the depression because of foreclosures. Large banks, insurance companies, and mortgage companies have taken over vast acreages not in contiguous tracts but in holdings scattered throughout the plantation belt. Appendix Table 8 indicates the holdings of these corporations in a number of sample counties in the States of North Carolina, Georgia, and Mississippi. This trend became serious in the boll weevil period, being well under way in the 1920's. The depression of the 1930's added still more foreclosures so that by 1934 an appreciable number of tracts were held by corporations.

The proportion which such corporation-held acreage formed of the total of all land in farms was as high as 20 percent in some counties. In the Georgia counties sampled the acreage held by corporations was about 11 percent of the land in farms, in North Carolina about 10 percent, and in Mississippi, 8.5 percent.

Method of Acquisition

Inefficiency in the operation of many plantations is not surprising when analysis is made of the means by which land was acquired. Of the 631 planters in the sample for whom data were available, 186 acquired their first tract by inheritance, 5 by marriage, 21 by foreclosure, 357 by purchase, and 62 by renting (Table 6).

Table 6—METHOD OF PLANTATION ACQUISITION
(Cotton Plantation Enumeration)

Method of Acquisition	First Tract		Additional Tracts	
	Number[a]	Percent	Number	Percent
Total plantations	631	100.0	220	100.0
Purchase	357	56.6	64	29.1
Inheritance	186	29.5	9	4.1
Marriage	5	0.8	-	-
Foreclosure	21	3.3	1	0.4
Renting	62	9.8	44	20.0
Combination	-	-	102	46.4

[a]Data not available for 15 plantations.

Over a third of the plantations surveyed contained tracts not in the original unit. The great majority of these additional tracts were bought or rented, or acquired by a combination of these two methods.

The date of acquisition of holdings indicates a considerable turnover in plantation ownership. Only 21 percent of the acreage surveyed had been acquired before 1910 and hence held at least 25 years. Forty-one percent had been acquired since 1925 and held less than 10 years and 21 percent had been held less than 5 years (Appendix Table 9).

ABSENTEE OWNERSHIP

The final stage in the decline of a plantation before its actual disintegration is when ownership is transferred to an absentee landlord. Foreclosure, inheritance, and speculative purchase often place the ownership of large tracts[2] in the hands of persons who are inexperienced in farming or occupied with other interests. Widows, other heirs, bankers, lawyers, merchants, and corporations become owners of plantations, but are unable to supervise their operation.

It is obvious that there are varying degrees to which a nonresident owner may be considered an absentee. He may hire an overseer to reside on and supervise the plantation, in which case the overseer, as agent of the owner, may be considered the resident operator. Overseer-operated plantations have been classified in this study with those of resident landlords. Also, landowners live at varying distances from their property and are restricted in varying degrees in the number of visits they can make to supervise operations. In order to allow somewhat for this variation, landlords were classified in this study as resident if they lived on the plantation, visited it daily, or employed an overseer; as semi-absentee if they lived within 10 miles and visited the place as much as once a week; and as absentee if they lived more than 10 miles from the plantation and visited it less than once a week.

According to this classification 85 percent of the plantations were operated by resident landlords, 9 percent by semi-absentee landlords, and 6 percent by absentee landlords (Appendix Table 10). The proportion of absentees varied somewhat in the various plantation areas, being highest in certain parts of the Black Belt and the Lower Mississippi Delta. These are also the areas in which a large proportion of the tenants are renters (Appendix Table 3), since the rental contract provides for very slight supervision by the landlord.

Another index of absenteeism is the extent to which landowners are partially dependent on other occupations. In this study it was found that 31 percent of all operators devoted more than one-fourth of their time to occupations other than farming (Table 7). This 31 percent is double the percentage of absentees

---

[2]It is probable that absenteeism is more prevalent in ownership of small tracts than in large since the investment in large tracts tends to create a pressure for use.

and semi-absentees combined, indicating that many landholders, though living nearby and frequently visiting their plantations, do not give their undivided attention to plantation operations. A landowner when he has some other occupation is most often a merchant.

Table 7—OPERATORS WITH OTHER OCCUPATIONS,[a] BY AREAS, 1934
(Cotton Plantation Enumeration)

Area	Total Operators	Operators with Other Occupations
Total	646	198
Atlantic Coast Plain	56	17
Upper Piedmont	40	9
Black Belt (A)[b]	112	33
Black Belt (B)[c]	99	34
Upper Delta	133	36
Lower Delta	50	15
Muscle Shoals	22	7
Interior Plain	30	11
Mississippi Bluffs	47	16
Red River	28	9
Arkansas River	29	11

[a]Enumerators were instructed to "enter other occupation from which operator derived an income in 1934 and to which he devoted more than one-fourth of his time."
[b]Cropper and other share tenant majority.
[c]Renter majority.

It is the absentee landlord who, through ignorance, laxity of supervision, or cupidity, most often allows the "mining" of the land and the loss of the productive top soil through erosion. It is on the absentee-owned plantations that fences and buildings most frequently fall into disrepair. It is these plantations which are least stable in a crisis. Since the owners of these tracts are most often holding the land for speculative or sentimental reasons, they do not "back up" the credit of their tenants to the same extent as do resident landlords. Hence, when a break in prices or a disaster like the boll weevil occurs and credit from merchants and bankers becomes tight, the tenants on these places are the first to find themselves without resources and are often forced either to move to the city or to become laborers or croppers on the more stable farms.

One of the former areas of extensive absenteeism which has been closely studied in relation to the boll weevil disaster is located in central Georgia between Macon, Atlanta, and Augusta.[3] According to these investigations the migration of tenants, loss of cotton production, and foreclosures on land and work stock were far greater in this region, and the pace of recovery after the boll weevil disorganization was slower than in other regions

---

[3]Raper, A. F., *Two Black Belt Counties: Changes in Rural Life since the Advent of the Boll Weevil in Greene and Macon Counties, Georgia*, unpublished Ph. D. dissertation, University of North Carolina, 1931, and Johnson, O. M. and Turner, Howard A., *The Old Plantation Piedmont Cotton Belt*, mimeographed report, Bureau of Agricultural Economics, 1930.

with less absenteeism. Table 8 indicates the drastic reduction in cotton production in counties with large absentee-owned plantations. Table 9 shows that the degree of reduction corresponded closely to the proportion of tenants.

Table 8—COTTON PRODUCTION UNDER THE WEEVIL 1921-1928 AS PERCENT OF COTTON PRODUCTION BEFORE THE WEEVIL 1905-1914
(59 Lower Piedmont, Georgia, Counties)

Year	Percent
Average - 1905-1914	100
1921	50
1922	34
1923	42
1924	57
1925	48
1926	62
1927	54
1928	56

Source: Johnson, O. M. and Turner, Howard A., *The Old Plantation Piedmont Cotton Belt*, Bureau of Agricultural Economics, 1930. p. 7.

Thus there have been two counterbalancing tendencies in land ownership. In the hands of efficient operators tracts have often been combined by purchase or lease. In the hands of absentee operators or inefficient farmers they have been loosely supervised, rented out in parcels, or actually broken up and sold in

Table 9—FIFTY-NINE COUNTIES OF THE COTTON-GROWING LOWER PIEDMONT (GEORGIA) GROUPED BY IMPORTANCE OF STANDING RENT, 1920, AND BY DECREASE IN AREA IN CROPS BETWEEN 1919 AND 1924

Decrease in Area in Crops between 1919 and 1924	Percentage of Standing Renters[a] among Tenants of Renter Status in 1920		
	59 or less	60 - 79	80 or more
Percent	Counties	Counties	Counties
19 or less	12	4	-
20 - 39	5	16	10
40 and more	-	1	11

[a]Tenants paying a stipulated amount of cotton as rental.
Source: Johnson, O. M. and Turner, Howard A., *The Old Plantation Piedmont Cotton Belt*, Bureau of Agricultural Economics, 1930. p. 12.

family-sized farms. As pointed out in the first part of the chapter, the size of tracts retained for taxation indicates little net change in concentration since 1910.

NEGRO OWNERSHIP

As noted in the first chapter, all Negroes were laborers immediately after the Civil War and the emergence of a landowning class was usually through the intermediate step of tenancy. This rise in status is reflected in Table 3 which shows 124,000 Negro landowners in 1910 in the 7 cotton States. The total for the whole South was over 200,000.

As with white owners, the disorganized conditions in cotton production from 1910 to 1930 caused a decline in number of Negro owners and proportion of Negro farms operated by owners. However, since both white owners and Negro owners decreased, the ratio of Negro to white owners was only slightly changed, 19 percent of the owners being Negro in 1910 and 18 percent in 1930 (Table 3).

As a rule, Negroes have been restricted in their opportunity to purchase land to the more undesirable sections. Just as white neighborhoods are recognized in cities so there are rural areas where Negro owners are not welcomed and white owners are reluctant to sell to Negroes. Thus Negro proprietorships have been acquired in outlying sections, on back roads, and on the poorer land.[4]

Few of the Negro proprietorships are of plantation size. The percentage of Negro landholdings in the various sized groups in

Table 10—SIZE OF PROPRIETORSHIPS BY COLOR OF OWNER IN 24 GEORGIA COUNTIES, 1934

Size of Proprietorship	Percent	
	White	Negro
Total	100	100
Less than 100 acres	47	74
100 to 260 acres	34	22
260 acres and over	19	4
Average size	213 acres	71 acres

Source: Tax digests.

Georgia[5] in 1934 as contrasted with white holdings is shown in Table 10; 74 percent of the Negroes owned less than 100 acres, 22 percent owned between 100 and 260 acres and only 4 percent owned 260 acres or more. It appears, therefore, that a negligible proportion of Negro landowners are in the "planter" class.

It is to be expected that Negro proprietorships would average much smaller than white, since the white proprietorships were large to begin with and Negroes owned no land in 1865. However, Negro proprietorships, like white, have been shrinking in size, owing to the faster increase in the number of small than of large holdings.[6] The average Negro holding in the Georgia counties in 1934 was 71 acres. The average white holding was 213 acres (Table 10).

[4]Raper, A. F., *op. cit.*

[5]Georgia is one of the few States where property holdings are segregated by color.

[6]Banks, E. M., *op. cit.*

Chapter III

PLANTATION ORGANIZATION AND MANAGEMENT

The visitor to well preserved *ante bellum* plantations is impressed with the variety of operations which were carried on and the division of labor practiced. The plantation included not only fields and barns but also spinning rooms, slaughter and storage houses, gins, grist mills, and other minor processing units. The plantation of today is not so nearly self-sufficing as were these old enterprises, but on large holdings there is still a marked division of labor even though production is more centered about the cash crop.

SIZE OF OPERATION

The very large plantations capable of a high degree of organization are gradually disappearing (Appendix Table 6). Table 11, classifying the plantations by crop acres, and Table 12, giving the number of resident families, indicate the small proportion of very large plantations. Only slightly over a tenth (69) of the plantations sampled in this study had 800 or more crop acres and about the same number (63) had 30 or more resident families. Furthermore, 55 of the 63 plantations with 30 or more families were concentrated in four areas—the Upper Delta, Red River, and Arkansas River areas, and the adjacent Bluffs section. In the Upper Piedmont and Muscle Shoals regions, about 90 percent of the plantations had less than 400 crop acres and less than 10 tenant families, and in the Black Belt and Coast Plain sections about three-fourths of the plantations had 10 tenants or less.

The definition of a plantation to include such a large proportion of small and medium-sized operations will not appeal to some who associate the word plantation with very large operations. Most descriptions have emphasized large-scale operations, both because these large operations are striking and because it is on these highly organized places for the most part that detailed records are obtainable. However, the distributions of the plantations in the special census enumeration of 1910 and of the sample of typical plantation areas in this study indicate that emphasis on such large-scale operations does not depict the true situation.

As in a manufacturing plant, the size of a plantation is the chief determinant of the degree of organization. A plantation with as much as 800 crop acres and 25 to 30 families is too large for the management and supervision of one man alone and functions must be delegated to managers, assistants, overseers, and riders. Such delegation of functions, however, represents merely a difference in degree of specialization and not an essential modification in the managerial function. The landlord or plantation operator has certain duties upon which the success of the operation depends and these duties are the same whether he performs them in person or whether he delegates them to subordinates. These duties include crop planning, finance of operations, management of labor and animal power, supervision of cultivation and harvest, marketing, and management of such processing enterprises as may be adjuncts to the plantation, such as commissary, gin, grist mill, blacksmith's shop, etc. In some instances the landlord also gives considerable aid in social and community problems of tenants, such as health, education, and religious life.

Table 11—ACRES IN CROPS ON PLANTATIONS, BY AREAS, 1934
(Cotton Plantation Enumeration)

Area	Total Plantations	Total Acres in Crops	Acres in Crops per Plantation	Number of Plantations with Specified Acres in Crops						
				Less than 200	200–400	400–600	600–800	800–1,000	1,000–1,200	1,200 and over
Total	646	248,513	385	241	213	81	42	29	15	25
Atlantic Coast Plain	56	16,473	294	20	25	5	6	–	–	–
Upper Piedmont	40	8,459	211	22	15	3	–	–	–	–
Black Belt (A)[a]	112	30,812	275	46	41	20	4	1	–	–
Black Belt (B)[b]	99	25,361	256	52	35	7	3	1	–	1
Upper Delta	133	74,873	563	23	41	21	13	17	10	8
Lower Delta	50	10,337	207	32	15	–	1	2	–	–
Muscle Shoals	22	4,943	225	13	8	–	1	–	–	–
Interior Plain	30	15,680	523	7	11	7	–	1	1	3
Mississippi Bluffs	47	17,769	378	21	9	5	6	3	1	2
Red River	28	14,865	531	4	11	3	5	2	–	3
Arkansas River	29	28,941	998	1	2	10	3	2	3	8

[a]Cropper and other share tenant majority.
[b]Renter majority.

The enterprises typically engaged in on a very large plantation are shown in Figure 10.

LANDLORD MANAGERIAL FUNCTIONS

The crop planning activities of the landlord are not so important on a cash crop plantation as they would be under diversified farming, but there is still the problem of how much to plant of the two major crops, cotton and corn, as a plantation total and as subdivided among the various tenants. This problem requires a knowledge of the family size and work habits of each individual tenant and of the character of the land to be assigned each tenant.

FIG. 10—ORGANIZATION OF ENTERPRISES ON THE LARGE AND CLOSELY SUPERVISED PLANTATION

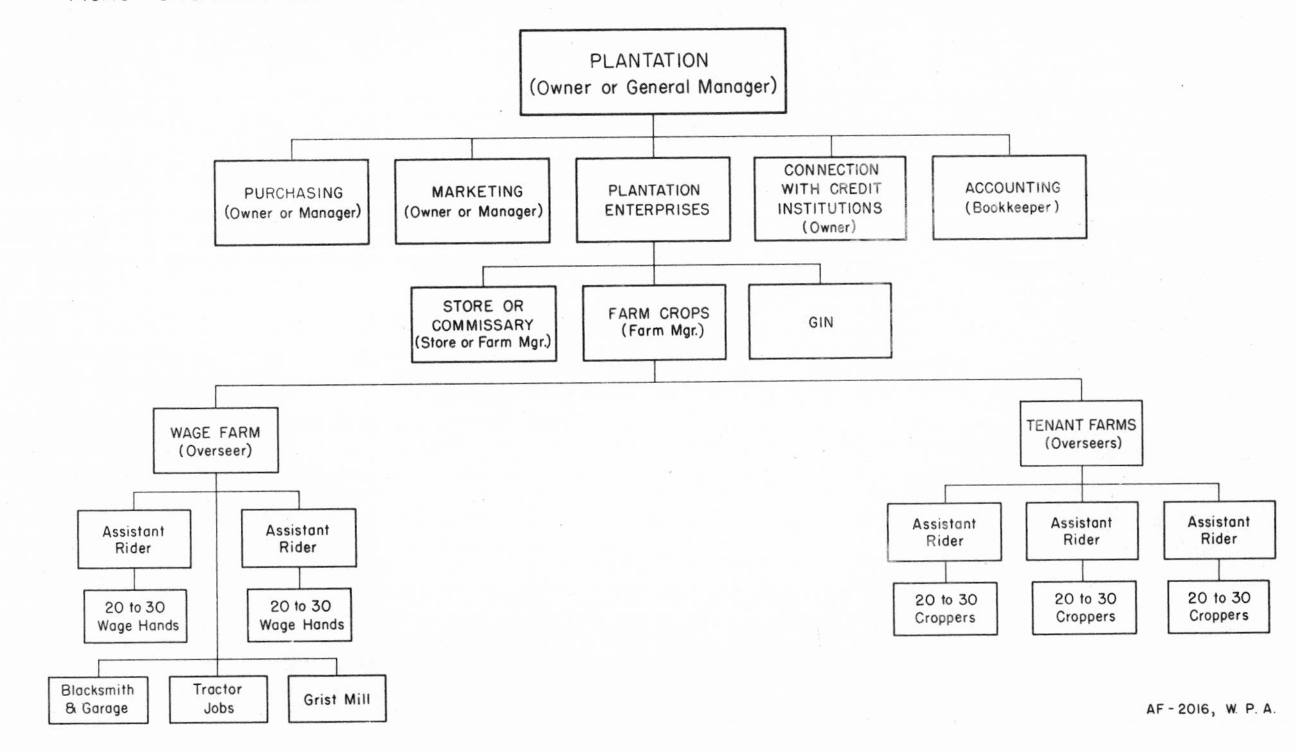

The task of financing operations is more fully discussed in chapter V. Appendix Table 11 indicates, however, that the planter's average investment of fixed capital, as estimated by representative planters, amounts to $28,694 per plantation (excluding operator's residence, gins, commissaries, and other non-agricultural equipment). The planter must also be continually on guard against deterioration of his drains, terraces, buildings, fences, and implements, and must be in a position to make repairs and replacements.

In addition to this outlay of fixed capital, the planter must be financially able to provide the working capital. Here the planter differs from the manufacturer. The well organized industrial enterprise has a fairly continuous outflow of working capital, and fairly continuous inflow of money from sales. The outflow of the plantation, however, begins in the early spring and there is comparatively little income until fall unless some of the previous year's crops have been held for sale. Thus, the planter must provide working capital without reimbursement for the major portion of the year, since the cropper or share tenant furnishes neither fixed nor working capital unless he owns his mule and plows. Theoretically the tenant furnishes his share of the seeds and fertilizer but in practice the landlord purchases the whole amount in advance and is reimbursed at the end of the season from the tenant's share of the crop.

Table 12—RESIDENT FAMILIES[a] ON PLANTATIONS, BY AREAS, 1934
(Cotton Plantation Enumeration)

Area	Total Plantations	Number of Plantations with Specified Number of Resident Families						Total Resident Families	Resident Families per Plantation
		5-10	10-15	15-20	20-25	25-30	30 & over		
Total: Number	646	358	108	48	40	29	63	9,215	14.3
Total: Percent	100.0	55.4	16.7	7.4	6.2	4.5	9.8		
Atlantic Coast Plain	56	40	11	3	2	-	-	482	8.6
Upper Piedmont	40	36	3	1	-	-	-	265	6.6
Black Belt (A)[b]	112	86	17	4	3	1	1	915	8.2
Black Belt (B)[c]	99	76	12	5	2	3	1	843	8.5
Upper Delta	133	28	29	15	12	15	34	3,115	23.4
Lower Delta	50	27	11	4	5	1	2	573	11.5
Muscle Shoals	22	19	3	-	-	-	-	148	6.7
Interior Plain	30	15	7	2	2	-	4	419	14.0
Mississippi Bluffs	47	23	6	4	5	4	5	725	15.4
Red River	28	7	7	5	3	1	5	554	19.8
Arkansas River	29	1	2	5	6	4	11	1,176	40.6

[a] Includes wage hands, croppers, share tenants, and renters (cash and standing)
[b] Cropper and other share tenant majority.
[c] Renter majority.

The planter is in the position of the type of manufacturer who pays wages for a number of months in advance of any income from sales. These payments are in the form of subsistence advances to tenants who must be fed and clothed on credit, the landlord to be reimbursed for these advances, and for his advances in seeds and fertilizer, when crops are sold. Theoretically, the tenant's income is a portion of the crop, collectible when the crop is marketed. In practice, the tenant collects this income in advance at the rate of from $5 to $20 per month, these subsistence advances amounting to an average of about $1,200 per plantation (Table 23, p. 63). At the beginning of the year, therefore, the average landlord must have, or be able to borrow, for subsistence advances and crop expenditures (Appendix Table 12), about $4,700 for total current operations.

This practice of subsistence advances is one of the chief trouble spots for the landlord. Tenants are given to understand at the beginning of the season that they are to have an advance of $5.00, $7.50, $10.00, $15.00, or $20.00 per month, depending upon how much the landlord feels he will be safe in advancing. Improvident families will use up their advance in the first 2 weeks of the month and try to overdraw during the last 2 weeks. The landlord must have continually in mind the needs of the family, the past record for "paying out", and the current condition of the tenant's crop in making decisions as to the extent of overdraft to allow. With a certain class of tenants the advance practice precipitates a continual struggle, the tenant attempting to get everything he can, the landlord attempting to advance as little as he can to keep the tenant at work.

Management of crops, labor, and animal power is the concern of the landlord throughout the spring and summer. The plantation day usually begins about sunrise and continues until dusk. The efficient landlord is in the field when the bell calls the tenants to work and visits each task at least once a day.

Some plantations keep the work animals in a central barn and assign them out as needed. Others regularly allot one or two mules to each tenant. The work animals on the plantations covered in this study averaged about one per resident family or 14 per plantation. This constitutes a considerable item of the landlord's investment which must be constantly safeguarded from deterioration through overwork, disease, or underfeeding.

Crop management also involves plans for fertilization and insect control, frequency and type of cultivation, and time of harvesting. Efficiency in all these matters rests far more on the judgment of the landlord than on that of the tenant.

The marketing function is also almost entirely in the hands of the landlord, except in the case of renters' crops. Having financed the operation in the spring and summer, the landlord controls the sale of the crop and division of the proceeds. His decision as to whether to sell immediately or to hold often makes a difference of several cents per pound on both his and the tenant's share of the crop.

RESETTLEMENT ADMINISTRATION

RESETTLEMENT ADMINISTRATION (MYDANS)

COMMISSARY AND MANAGER'S HOME ON LARGE PLANTATION IN MISSISSIPPI

NON-AGRICULTURAL ENTERPRISES

It is beyond the scope of this study to comment in detail on the non-agricultural enterprises of the larger plantations, such as gins, shops, mills, and commissaries. These are merely adjuncts to the production and marketing of the cash crop and as such are operated only by part of the plantations.

The commissary is one of the most criticized plantation features but may, if fairly administered, be of advantage to the tenant. Usually the commissary is introduced by the operator so that he, rather than the supply merchant, may control the expenditures for subsistence and keep these amounts within the limits of the tenant's ability to produce. The landlord also gets the advantage of wholesale prices with discounts if he is able to finance purchases in cash.

The advantage, or disadvantage, of this practice to the tenant depends entirely upon the extent to which the landlord passes the economies of wholesale buying on to the purchaser and the extent to which the landlord merely substitutes himself for the exploitative merchant. Many examples of both types could be produced. In the present study more than a fourth of the plantations had commissaries, 15 percent having commissaries whose use by the tenants was compulsory and 11 percent commissaries whose use was optional (Appendix Table 13).

For performing these varied services, the average landlord makes little more than the cost would be of hiring managers and overseers to do them. Calculated on the basis of salaries paid in 1920 on large plantations, Brannen concluded that the cost of hired supervision was $1.80 per acre[1] on cotton plantations. According to the present study the landlord profit (after deducting interest on his capital) averaged only $2.01 per crop acre.[2]

SOCIAL CONTRIBUTIONS

Having established and perpetuated a paternalistic relation to tenants and having taken the responsibility for close supervision not only of agricultural operations but also of family expenditures, the landlord is also often called upon for services of a social nature, for the large plantation is a social as well as an economic organism and the matrix of a number of plantations often constitutes or dominates the larger unit of civil government in the locality.

Among efficient landlords, tenant health is one of the major considerations and doctors' bills are paid by the landlord and

---

[1]Brannen, C. O., *Relation of Land Tenure to Plantation Organization*, U. S. Department of Agriculture, Bulletin 1269, pp. 17-19.

[2]See chapter VI.

charged against the tenant crop. Those tenants who have a landlord who will "stand for" their bills are far more likely to get physicians' services than are the general run of tenants. On some plantations socialized medicine is approximated. The landlord pays a flat rate to a doctor who agrees to serve all the tenants for a year, and this charge is distributed on a per visit basis. On plantations where medical contributions or advances are made by the management these average about $40 per plantation (Appendix Table 14-A).

Landlords and managers are also expected to "stand for" their tenants in minor legal difficulties such as may grow out of gambling games, altercations, and traffic infractions. This function is, of course, not exercised indiscriminately. A good worker will, in all probability, be "gotten off" and a drone left in the hands of the law. In past decades, the sheriff seldom went on large plantations, minor discipline being one of the manager's undisputed prerogatives. The broad leather strap was the principal instrument of discipline. These practices of plantation discipline have passed, but the landlord assumes responsibility for such tenants as are arrested for minor offenses, especially during the busy season. In the present study 11 percent of the landlords had, in the year 1934, acted as parole sponsor for tenants and 21 percent had paid fines (Appendix Table 14-B).

Use of plantation animals for social or personal purposes is also one of the plantation contributions. Three-fourths of the plantations studied allowed the use of their animals for trips to town and on Sundays, but of these more than one-fourth did not allow such use as often as once a month. Thus, a large proportion of the tenants who did not own work animals either had no means of transportation or had such means available less than once a month (Appendix Table 14-B).

Landlords are also frequently expected to contribute to plantation social life through aid to churches, schools, and entertainments. The present study revealed that direct contributions to schools were relatively small although the use of land for school buildings was frequently permitted. Planters interviewed reported an average annual contribution to tenants' churches of approximately $13 and to tenants' entertainments of $6 (Appendix Table 14-A). In addition, it goes without saying that plantation waters are open to tenants' fishing, and plantation rabbits and quail are theirs for the taking. Usually the landlord's contribution of supplies for entertainments such as fish fries, barbecues, and dances is more substantial than his cash contribution.

The contribution of the landlord to plantation efficiency may be summarized as that of the pocketbook and brain. The contribution of the tenant is largely that of supervised brawn.

Landlords vary widely in their capability of performing these functions efficiently. Some prefer to stay in town rather than ride over their land. Others work very energetically at their job, thereby contributing materially to their own fortune and to that of their tenants. It is clear that the efficient landlord is not only a capitalist, but also an agronomist, a diplomat, a capable manager, and occasionally a veterinarian and social arbiter.

Tenants have traditionally depended on landlords for services such as those described in this chapter and any plan for replacing the plantation organization with other forms of tenure or small ownership must take into consideration the reality of the managerial function and the practical necessity for supervision of a plantation. It must either provide a similar management until tenants outgrow its need or, through intensive education, train the tenants to perform these duties efficiently.

Chapter IV

THE ONE-CROP SYSTEM

Although the plantation is excessively devoted to the production of money crops it must also furnish fuel for the tenants and feed for the work animals. Hence most tracts include woodland and pasture as well as crop land for feed purposes. Also, many plantations, especially since the A.A.A. crop reduction program, contain bodies of idle or fallow land. Each plantation has a reserve of land which can be brought into cultivation or left idle according to price prospects. This is one of the reasons why cotton production fluctuates so violently.

LAND USE IN PLANTATIONS

The land distribution in the plantations studied differs from that of the total farm land in the plantation States in

Table 13—USE OF LAND IN 646 PLANTATIONS AND OF TOTAL LAND IN FARMS IN SEVEN SOUTHEASTERN COTTON STATES,[a] 1934

Land Use	Percent Distribution	
	646 Plantations	Total 7 States
Total	100.0	100.0
Crops	42.4	35.3
Pasture	17.8	24.5
Woods	23.6	28.5
Idle	7.0	6.3
Waste	9.2	5.4

[a] Alabama, Arkansas, Georgia, Louisiana, Mississippi, North Carolina, and South Carolina.

Source: Cotton Plantation enumeration and *United States Census of Agriculture, 1930*.

that the plantations have a considerably higher percentage of crop land and a correspondingly lower percentage of land in woods and pasture (Table 13).

The percentage of crop land in the total acreage was highest in the plantations of the Upper Delta of the Mississippi and the bottom and bluff land of its tributaries, where it amounted to more than half the total acreage.

In the Upper Piedmont section crop land was just under 50 percent of the total. The sections of high absentee ownership (Black Belt renter counties and Lower Delta) show a low proportion of their land in crops and a relatively high proportion in woods and pasture (Appendix Table 15).

The typical plantation of this inquiry is 907 acres in extent with 385 acres in crops, 214 in woods, 162 in pasture, 63 idle, and 83 waste (Appendix Table 15). This average of 907 total acres is high, owing to the effect of the comparatively large number of plantations of 1,500 acres or more. Fifty percent of the tracts are between 250 and 750 acres in size (Table 14).

After the landlord has determined the number of families he can finance and the acreage which he can conveniently and economically plant to cotton, he allows the balance of his land to grow up to woods and so-called pasture if it is not too severely eroded. A considerable part of the idle land and some of the woods and pasture could, if necessary, be converted to additional crop acreage.

Table 14—ACRES OPERATED[a] ON PLANTATIONS, 1934
(Cotton Plantation Enumeration)

Acres Operated	Plantations	
	Number	Percent
Total	646	100.0
Less than 250 acres	71	11.0
250 to 500 acres	173	26.8
500 to 750 acres	149	23.1
750 to 1,000 acres	70	10.8
1,000 to 1,250 acres	55	8.5
1,250 to 1,500 acres	35	5.4
1,500 acres and over	93	14.4

[a]Owned and rented; includes crop land, tillable land lying idle, pasture land, woods not pastured, and waste land.

CROP ACREAGE PER FAMILY

Under the present system of cotton-corn farming, plantation land is used more intensively than land in most other sections of the country. The average of 25 crop acres per family (Appendix Table 16) is far less than that in the Middle West or Far West but larger than in the self-sufficing farm regions such as the Appalachians. It is also a larger acreage per family than in the trucking and high price crop specialty regions.

Croppers who specialize in cotton show an average of 20 acres in crops, one-fifth less than the average crop acreage for all plantation families. Other share tenants and renters, who produce more food and feed for home use, average 25 acres, while wage hand families, cultivating the unshared landlord crop, average 45 acres (Appendix Table 16).

It is also apparent that the sections specializing most heavily in cotton (the Delta and the Arkansas River bottom) assign smaller acreages per family, the average cropper acreage in the Upper Delta being 17, in the Lower Delta 15, and in the Arkansas River valley, 14 per family. That these smaller cropper acreages are often accompanied by larger acreages per wage hand family indicates a different plantation organization in these sections. Here the croppers concentrate almost entirely on cotton and the food and feed crops are grown by the landlord with wage labor.

The cropping arrangement of the Cotton South is usually spoken of as the one-crop system but it is in fact a two-crop system, cotton for the cash crop and corn for food and animal feed. Almost as much acreage is planted to corn as to cotton, but corn is a minor item in the cash transactions of the farm, since it is produced largely for home use. In fact, many plantations do not plant enough corn to feed the work animals. On the plantations studied, 4 percent of the total expenditure was for feed although feed could easily have been grown (Appendix Table 12).

Money-crop farming was highly developed on the large plantations of the South by the close of the Civil War, but in the subsequent expansion of the demand for cotton this system fastened its hold even more firmly on the section. The better managed slave plantations produced in addition to cotton, considerable quantities of foodstuffs, many slaves being required to cultivate gardens of their own. But plantations have become less self-sufficing during recent decades. Total cotton acreage in the United States expanded from 12 million acres in 1875 to 45 million in 1930 while livestock and food production suffered a relative decline.[1]

The generation after the Civil War grew up with little knowledge of farming except the minimum necessary for growing money crops. Gardens largely disappeared and the habits of caring for livestock were often lost. The South, though a section suited by soil and climate to the culture of a great variety of food crops, became a heavy purchaser of foods from other sections. Though a section with vast areas suited to stock production, it became dependent for its mules and dairy products upon farmers elsewhere. The tenant, the landlord, the merchant, and the banker all conspired, because of short-sighted self-interest, to expand the money crop, with the result that the South depleted the fertility of vast tracts and allowed them to erode. It became enmeshed in a vicious tenant system, and dependent upon ruinous credit machinery.

The plantations studied were selected from cotton counties,[2] but manifested a wide range in proportion of crop acreage planted to cotton. A few had less than 20 percent of their acreage in cotton in 1934 and a few had more than 70 percent of their acreage in cotton, but the great majority operating under the

---

[1] *Yearbooks*, U. S. Department of Agriculture.

[2] A cotton county was defined as a county in which 40 percent or more of the gross farm income was from cotton.

crop reduction program devoted from 30 to 60 percent of their acreage to cotton (Appendix Table 35).[3]

TRENDS IN COTTON CULTURE

While the increase in cotton acreage was from 12 to 45 million acres in 55 years and the production increase was from 5 to 15 million bales, these increases were not uniformly distributed.[4] Practically all cotton was concentrated in the old Southeast in 1870, but by the turn of the century a third of the acreage was in the two States of Texas and Oklahoma and by 1930 half the acreage was in these two States. The tremendous relative loss of the old Southeast is shown if the heavy cotton producing States are divided into three groups and the average acreage of the years 1906-1910 and 1926-1930 compared (Table 15).

Table 15—AVERAGE ACREAGE IN COTTON, BY GROUPS OF STATES, 1906-1910 AND 1926-1930

States	Acres (in thousands)		
	1906-1910	1926-1930	Percent Change
Texas, and Oklahoma	11,348	22,547	+ 99
Mississippi, Arkansas, and Louisiana	6,795	9,510	+ 40
Alabama, Georgia, North Carolina, and South Carolina	12,094	11,476	- 5

Source: *Yearbooks*, U. S. Department of Agriculture.

The reduction in the Southeast and shift to the Southwest largely resulted from boll weevil damage in the Southeast. The incidence of heavy damage by the boll weevil was not a sudden catastrophe affecting the whole South simultaneously or equally. Rather it was a wave whose crest moved from West to East at from 40 to 160 miles per year. The effect of the weevil was varied in the States ahead of its march, in the States recently invaded, and in the States where its maximum damage was past.

The passage of the weevil in Texas and Oklahoma occurred before the great expansion of acreage in these States, and when the expansion took place it more than compensated for the weevil damage. From these States eastward, however, the process was as follows:

Ahead of the weevil, States received price benefits from the curtailed crop of their western neighbors and this led to increases in acreage and intensified fertilization. Hence, the weevil was preceded by increased production and increased value of cotton crops.

The desire to plant "one more big crop" led farmers to continue past the danger point with the result that in the first

---

[3]See chapter VI.

[4]"Statistics of Cotton", *Yearbooks*, U. S. Department of Agriculture.

year of heavy infestation crops upon which heavy expenditures had been made were destroyed. This loss completely disorganized credit, ruined tenants and landlords alike, and resulted in greatly curtailed acreage, increase in share-cropping, decrease in ownership and cash renting, and reduction of credits. After the period of heavy damage there was a gradual climb upward as farmers learned to cultivate cotton under weevil conditions, recouped their losses, and diversified to a greater extent.

The resultant changes in acreage, production, value, and tenancy as they appear in the census statistics depend upon which stage in this cycle the State happened to be in at the time of the enumeration.

If 1910 is taken as a base line, it is found that by that date the weevil had crossed the Mississippi River hardly in sufficient numbers to be felt. Louisiana and Arkansas were recovering and production was expanding in Alabama, Mississippi, Georgia, North Carolina, and South Carolina. Soon after 1910, production in Mississippi fell off sharply. In the period 1915-1920 Alabama had its big drop in production and Georgia began to decline. From 1920 to 1930 Georgia and South Carolina were the great sufferers with some damage in North Carolina while Mississippi and Alabama recovered completely and Louisiana, Arkansas, Texas, and Oklahoma continued to forge ahead.

From 1923 the acreage in cotton began a steady march upward, reaching a new high of 47 million acres 3 years later in 1926. The average for 1921-1925 was over 37.5 million acres; for 1926-1930 it was 44.5 million acres.[5] This phenomenal increase in acreage was the result of two factors. The expansion into the level lands of the southwestern prairies, high plains, and red plains, given its first impetus by the boll weevil, continued at an increasing rate. At the same time richer areas in the Southeast were recovering from 1925-1932 some of the ground lost from 1920 to 1924. Alabama, Georgia, and South Carolina had failed to reach their maximum acreage before weevil infestation. Arkansas, Louisiana, and Mississippi, on the other hand, continued to expand their acreage after 1923. It is notable that the northern border States, Tennessee, North Carolina, and Virginia, reached their greatest acreage in the period of greatest weevil damage in 1923-1924 and their acreage has receded with the recovery of other areas.

The extent of the disorganization of cotton production prior to 1930 was largely a function not of production alone but of the value of the crop. Although Mississippi production declined sharply from 1910 to 1920, rises in price meant an actual increase in the value of the crop. Mississippi was, therefore,

---

[5] *Yearbooks*, U. S. Department of Agriculture.

only temporarily disturbed by the weevil. To a lesser extent the same condition was true in Alabama. By the time of the short crops in Georgia and South Carolina, however, the increase in production to the westward and the financial difficulties of the early 1920's caused short crops and low prices to coincide, with the result that the disorganization in these two States from 1910 to 1925 was drastic. This was not registered to any marked extent in the 1920 Census but showed up sharply in the 1930 enumeration.

Thus, even before the crash of 1929, parts of the Old South, especially of the Southeast, were sharply depressed. This depression in cotton had already resulted in a relative decline in land ownership and loss of work stock by tenants. From 1915 to 1928-1929 hundreds of thousands of Negro tenants and some white tenants moved out of cotton counties to industrial cities. Wher further disorganization of prices and finances occurred in 1928-1929 hundreds of additional planters had to discontinue operations and thousands of tenants who had no one to feed them were forced out of agriculture.[6]

The accumulated overproduction of the early 1930's may, therefore, be described as the result of the efforts of the States east of the Mississippi to regain their former place in the cotton picture while the States west of the Mississippi continued to expand their acreage.

The very large crop of 16 million bales in 1931 was added to a carry-over of 10 million bales from previous crops. This supply of 26 million bales was more than twice the usual annual consumption requirements. At the same time world market losses were restricting exports and domestic consumption was falling off.

The result was that in 1932 the price of the staple fell to a new low for recent years of six cents per pound. This was materially below the cost of production for most farmers and many marginal producers were forced into bankruptcy.

The relatively more frequent and more violent depressions in cotton price than in the prices of agricultural commodities as a whole account for the greater financial difficulties of the South as compared with other agricultural areas. From Appendix Table 17 it is apparent that there was a violent depression in cotton in 1926 and 1927 which hardly affected other commodities. The ratio of cotton price to cost of living fell from 94 in 1929 to 44 in 1932, while the general index of agricultural commodities fell only from 95 to 61, a decline of more than 50 percent for cotton in comparison with about 35 percent for all crops (Appendix Table 17 and Figure 11).

---

[6] See chapters V and X.

The relative severity of the depression in the cotton States and in the rural United States as a whole is also indicated by the change in net time and demand deposits of banks in towns of less than 15,000 population (Appendix Table 18 and Figure 12). Taking 1929 deposits as 100 percent, the lowest level for all rural banks was 52 percent in 1934 while the lowest level for southeastern banks was 31 percent in 1933. The beginning of recovery in the South was a year ahead of that in the rest of the Nation. From the low point of 31 percent in 1933 the index of cotton States' deposits climbed to 38 in 1934 and 53 in 1935, while the low point for the whole United States was in 1934 with an upturn in 1935. From January 1934 to January 1935 the index moved up from 52 to 62.

The A.A.A. years, 1933, 1934, and 1935, brought a material improvement in the condition of the cotton farmer. Aside from any effect of the payment of rental benefits, the improvement in price revolutionized the position of the grower. This is apparent from the following comparison of the receipts of a producer of 10 bales in 1932 with his receipts after reducing his output 40 percent in 1935 in accordance with A.A.A. requirements.

Year	Number of Bales	Price per Pound	Value per Bale	Total Value
1932	10	6¢	$30	$300
1935	6	12¢	$60	$360

Not only was his gross income 44 percent higher but his expenses were less in producing the six bales and his idle land was rented by the government.

Fertilizer Consumption

Cotton and corn, the mainstays of the plantation, are soil-exhausting crops and have been cultivated on plantation lands so long that much of the natural fertility has been mined out of the land. Southern agriculture is more dependent upon fertilizer to supplement the plant food in the soil than any other section. Figure 13 shows the fertilizer consumption by States.

The present study indicates that, of the average crop expenditure of $3,472 per plantation in 1934, $336 or 10 percent, was for fertilizer (Appendix Table 12). Since a large proportion of the expenditure is credit purchase on which high rates of interest are paid, fertilizer and interest on fertilizer bills are large items in the plantation budget. On another basis, this amounts to an expenditure of almost a dollar (87 cents) per crop acre (Table 11). When it is considered that

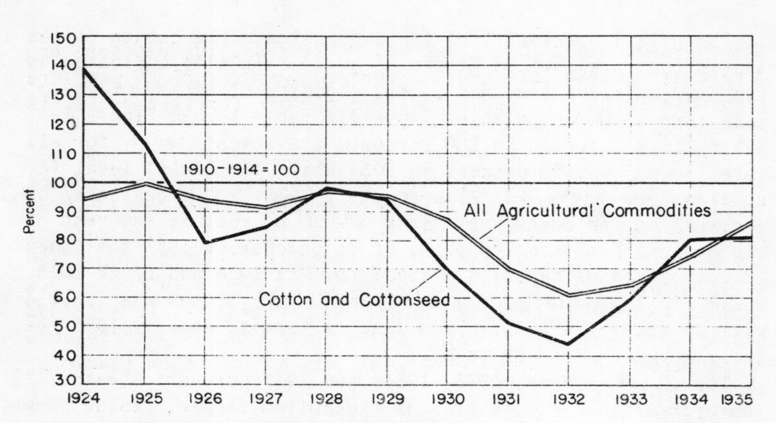

FIG. 11 - RATIO OF PRICES RECEIVED FOR COTTON AND COTTONSEED AND FOR ALL AGRICULTURAL COMMODITIES TO PRICES PAID FOR COMMODITIES BOUGHT 1924-1935

Source: Bureau of Agricultural Economics

AF-1425, W.P.A.

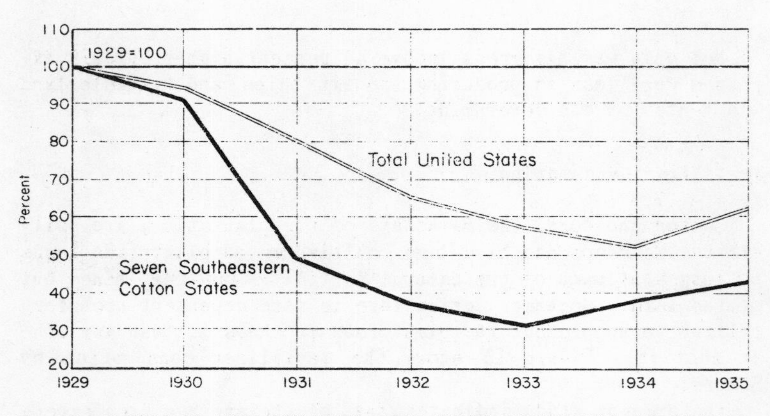

FIG. 12- AVERAGE NET DEMAND AND TIME DEPOSITS IN BANKS LOCATED IN TOWNS OF LESS THAN 15,000 POPULATION 1929-1935

Source. Annual Reports of Federal Reserve Board

AF-1423, W.P.A.

the operator's net labor income, after deducting expenses and 6 percent on the investment, averages just over $2 per acre (Appendix Table 37), the importance of the fertilizer bill is emphasized.

There is considerable variation in the extent to which fertilizer is needed on different types of land, the percentage of total expenditure for fertilizer varying from 2 percent of total plantation expenditures in the Red and Arkansas River bottoms and 4 percent in the Mississippi Delta to 33 percent in the sandy Coast Plain and 32 percent in the Upper Piedmont. This amounts to over $3 per crop acre in the Coast Plain area and nearly $2 in the Upper Piedmont area, resulting in a tremendous differential in favor of the Delta and Bottom Lands (Table 11 and Appendix Table 12).

SOIL EROSION

SOIL CONSERVATION SERVICE

SOIL EROSION TYPICAL OF ROLLING LAND UNDER CLEAN CULTIVATION

The traveler through the older lands of the Southeast receives two predominant impressions from the landscape—red hills and muddy streams, both of which are products of soil erosion. Outside of the level Delta and Lower Coastal Plain areas, the cotton country is rolling and subject to erosion because of the topography and the physical character of the soil. The tremendous tonnage of topsoil which washes off the surface of southern lands not only depletes the farm but also contributes to serious engineering problems. Reservoirs are silted, navigation channels fill up, and flood control is made tremendously more difficult.

Rudimentary erosion control was inaugurated years ago on the rolling lands by placing the furrows along contours rather than in a straight line and by terracing along contours. This device extended the cultivable acreage of the South by millions of acres. It is not sufficient, however, to control the loss of topsoil. The practices advocated by the Soil Erosion Service are contour farming and strip cropping, combined with sound crop rotation, including the use of winter cover crops; sound pasture management, including establishment and maintenance of good turf, regulated grazing, use of contour furrows, water diversion, etc.; and proper woodland management. Such changes in southern agriculture are difficult to accomplish under a tenant system on large tracts geared to cotton and corn production. On small one-man farms they are feasible and in the long run profitable, but where debts must be met from cash income and tenants must be financed, such diversified agriculture does not produce cash or pile up profits for an operator whose profit comes from the production of a number of tenants.

The result is that the Southeast is one of the major erosion problem areas of the Nation. Figure 14 indicates the incidence of erosion and its seriousness in the southeastern region. It will be noted that the most heavily eroded area follows the eastern plantation belt, *i.e.*, the Upper Atlantic Coast Plain and Lower Piedmont sections which extend southwest paralleling the Appalachian Mountain Ridges.

A reconnaisance survey by the Soil Erosion Service[7] of 7 southeastern States shows 10,900,000 acres essentially destroyed for further tillage. Numbers of other farms will follow the same road to destruction within a short time unless there is a radical change in the cropping system.

Erosion and other causes of low soil productivity are summarized in the statistics of the National Resources Board on submarginal land.[8] In mapping the land for retirement from arable farming this report includes 1,600,000 cotton acres in the retirement areas. This is 10 percent of all land proposed for retirement and 3.7 percent of the total cotton land.

Thus an appreciable proportion of the cotton production is carried on below the "margin" and much more is almost a marginal activity. In years of good yield and prices small profits are realized on this land, but in bad years losses are severe.

---

[7]Unpublished.

[8]*National Resources Board Report, December 1, 1934*, pp. 160-184.

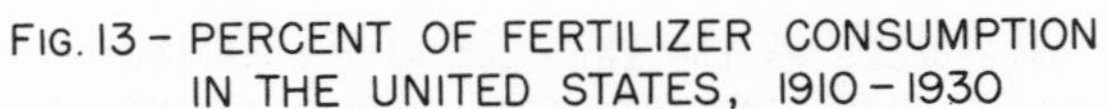

FIG. 13 – PERCENT OF FERTILIZER CONSUMPTION IN THE UNITED STATES, 1910 – 1930

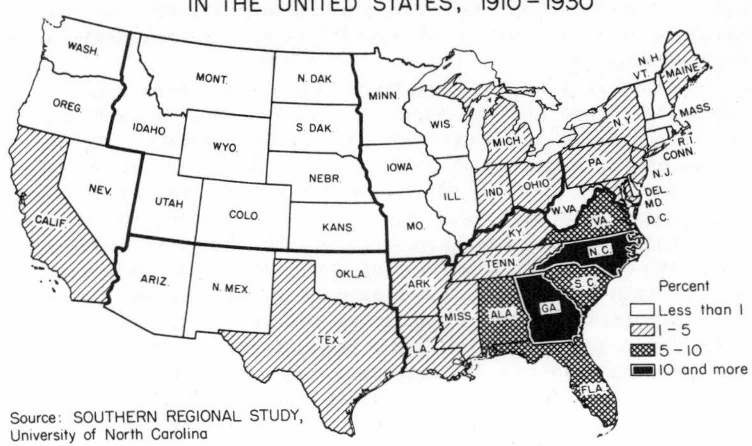
WASH.
OREG.
CALIF.
NEV.
IDAHO
MONT.
WYO.
UTAH
COLO.
ARIZ.
N. MEX.
N. DAK.
S. DAK.
NEBR.
KANS.
OKLA.
TEX.
MINN.
IOWA
MO.
ARK.
LA.
WIS.
ILL.
MICH.
IND.
OHIO
KY.
TENN.
MISS.
ALA.
GA.
FLA.
S. C.
N. C.
VA.
W. VA.
PA.
N. Y.
N. J.
DEL.
MD.
D. C.
CONN.
R. I.
MASS.
VT.
N. H.
MAINE
Percent
Less than 1
1 – 5
5 – 10
10 and more
Source: SOUTHERN REGIONAL STUDY,
University of North Carolina
AF – 2019, W. P. A.

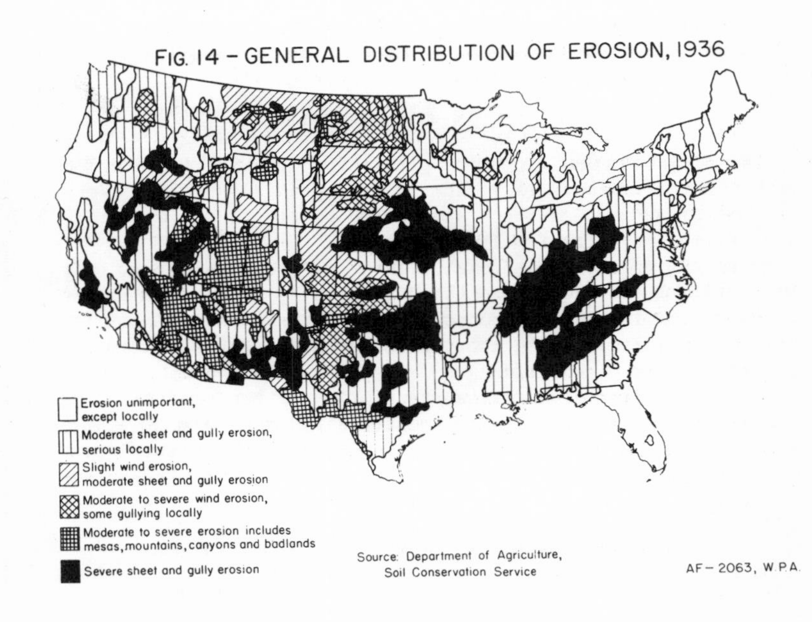
FIG. 14 – GENERAL DISTRIBUTION OF EROSION, 1936
Erosion unimportant, except locally
Moderate sheet and gully erosion, serious locally
Slight wind erosion, moderate sheet and gully erosion
Moderate to severe wind erosion, some gullying locally
Moderate to severe erosion includes mesas, mountains, canyons and badlands
Severe sheet and gully erosion
Source: Department of Agriculture, Soil Conservation Service
AF – 2063, W. P. A.

The report points out further that:

> Extensive areas of submarginal land, however, are devoted to commercial types of farming. In such areas there is a tendency to incur excessive indebtedness during temporary times of good prices. The volume of such debt fluctuates widely between periods of prosperity and depression. The characteristic course of development during periods of rising prices is that of a growing percentage of properties mortgaged and a rapidly accumulating volume of debt until the proportion of total value represented by indebtedness results in narrow equities by owners. A collapse of prices discloses a general absence of reserves and is quickly followed by wholesale delinquencies, foreclosures, and surrender of titles because of debt pressure. Within a few years the proportion of owner-operated farms may become greatly reduced as vast tracts of land are foreclosed and much of it is abandoned. Such submarginal areas are typically far from market, and, since transportation costs are relatively constant, farm incomes and value of land are reflected in extreme fluctuations.
>
> Behavior of this course of events....*is*[9] illustrated by certain parts of the wheat plains of the West and of the cotton-growing lands of the Southeast and other territories in which temporary periods of high prices have induced unduly expanded production and increase of equipment....[10]

PRODUCTION FOR HOME USE

The slight extent to which plantation economy is concerned with animal husbandry is indicated by Appendix Table 19. Very few cows, pigs, and chickens are raised by the landlord, hardly more than enough for family consumption.

Only 15 percent of the total plantation gross income in 1934 for landlord and tenants combined (excluding rent of land let to renters and home use production of renters) was for home use and this amounted to only about $100 per family.[11] This shortage of home use production puts the South in the position of purchaser of many commodities which it could produce. The following tabulation of Alabama production indicates how far short that State falls of producing its subsistence (Table 16).

---

[9] Italics ours.

[10] *National Resources Board Report, December 1, 1934*, pp. 179-180.

[11] See chapter VI.

DIVERSIFICATION OF CROPS

For the past 20 years continuous pressure has been exerted on Cotton Belt farmers to diversify their operations, but without marked results. One of the central features of the farm demonstration programs has been the promotion of diversification.

Table 16—PERCENT OF NEEDS PRODUCED IN THE STATE OF CERTAIN ALABAMA FARM PRODUCTS, 1928

Commodity	Percent of Needs Produced
Corn	40
Hay	30
Oats	4
Meat	73
Milk	63
Eggs	137
Potatoes	89
Syrup	110
Apples	24
Peaches	95
Vegetables for sale (acreage)	120
Vegetables for home consumption	40
Cotton[a]	40

[a]Consumed in Alabama mills, not necessarily by Alabama people.

Source: Data compiled by the State Statistician of Alabama, Quoted by Rupert Vance, in *The Negro Agricultural Worker*, unpublished report for the Committee on Negroes and Economic Reconstruction.

Occasionally this has borne fruit in specific localities such as Turner County, Georgia, where the "cow, hog, and hen" program has markedly increased self-sufficiency, and Colquitt County, Georgia, where a packing plant provides an outlet for livestock products. In a few Mississippi areas the establishment of processing plants has made dairy products profitable and in

Table 17—LIVESTOCK ON FARMS, IN SEVEN SOUTHEASTERN COTTON STATES,[a] 1930 AND 1935

Livestock	1935	1930
Cattle[b]	6,964,237	4,194,090
Cows and heifers 2 years old and over	3,719,115	2,441,916
Hogs and pigs	6,208,626	5,767,070

[a]Alabama, Arkansas, Georgia, Louisiana, Mississippi, North Carolina, and South Carolina.

[b]Excludes animals under three months, April 1, 1930.

Source: *United States Census of Agriculture, 1935*, preliminary figures.

a few areas specialty crops, such as tomatoes, peppers, small fruits, or orchard crops, have been cultivated on a limited scale.

The boll weevil provided another short-lived impetus to diversify. In an excess of enthusiasm over the expected increase in prosperity from enforced diversification, a southern Alabama town erected a monument to the boll weevil soon after its onset. Ten years later, however, this county was cultivating almost as much cotton as before the advent of the weevil.

Again the World War, with pressure from food administrators provided some impetus for diversification but at the same time cotton prices were sufficiently tempting to more or less nullify this influence.

The latest pressure to diversify has come from the enforced cotton reduction program of the A.A.A. The results of the 1935 Census of Agriculture indicate that probably more diversification has been induced between 1933 and 1935 than in any other period of the South's history. Table 17 indicates the remarkable increases in livestock in the 7 southeastern cotton States. Supplementing this is the fact that from 1930 to 1935 there was a marked increase in acreage in pasture in southeastern farms. There has also been a pronounced increase in acreage devoted to corn, wheat, peanuts, and hay crops, as indicated by the 1935 Census.

On the basis of past experience, however, there is no guarantee that this latest trend to diversification will remain permanent if restrictions on cotton production are removed and prices return to normal levels. The boll weevil imposed a restriction on cotton production in the 1910's and 1920's as drastic as that of the A.A.A. but no permanent diversification resulted.

Diversification of southern agriculture on a permanent basis rests on two necessities. The first is an increase of production for home consumption, which is predicated on a high standard of living for workers in agriculture. The present share tenant system is a major stumbling block to the realization of this condition, since the landlord is not financially interested in tenant home use production. The second condition, which has been met for only a few commodities in limited areas, is the development of markets where other commodities such as peanuts, sweet potatoes, soy beans, and the variety of crops which will thrive in southern soils and climates may stand alongside of cotton and tobacco as producers of cash income.

Chapter V

CREDIT

Except after an unusually prosperous year the majority of the plantation population operates on borrowed money for a number of months. Landlords of plantation operations, as has been pointed out, need on the average $3,500 during the crop season to pay current crop expenses (Appendix Table 12), an average of $1,200 to feed their tenants until the crop can be sold, if subsistence advances are made, and $150 to pay wages. Tenants, even if not in debt from previous years, usually exhaust their slender resources by January or February and begin to live on credit until the following harvest.

Thus, both landlord and tenant have two types of debt. The first is the result of past deficits or purchases. In the case of the landlord this long term debt is usually secured by a mortgage or bank note. In the case of the tenant, accumulated debts are usually in the form of an unsecured account with the landlord or merchant which stands as a lien against future production. In a few instances, the tenant's debts are secured by a chattel mortgage on livestock and implements. The second type of debt is the short term loan to meet the expenses of the current crop. In this case the security offered by both landlord and tenant is usually a lien against the crop.

The growing crop is such uncertain security for these current loans that interest rates are ruinously high. Few businesses could operate profitably with interest charges amounting to such a substantial item of expense as they do in the cotton farmer's budget. The Cotton Belt lender as well as the borrower is often wiped out financially in poor years. Having borrowed in advance to finance production, in years of poor yields or low prices the tenant is often unable to discharge his debt to the landlord. The landlord, in turn, does not net a sufficient income to repay the loans for the tenant's share of expenses as well as for his own. Thus, the merchant, unable to collect his accounts, has a large share of his assets frozen, and if pressed by his creditors he may become insolvent. Banks, which finance merchants and landlords, in turn are unable to collect. The whole economy of the Cotton Belt is, therefore, based on a gamble as to the yield and price of the crop, a game played with borrowed stakes.

LANDLORDS' LONG TERM INDEBTEDNESS

Twenty-five years ago short term credit constituted a major problem to farm owners as well as to tenants in the Cotton Area. Since then long term credit, chiefly in the form of mortgages, has become increasingly important.

Appendix Table 20 shows the 1934 long term indebtedness of the landlords on the plantations sampled. Almost 90 percent of this fixed indebtedness was in the form of mortgages. In 1934 nearly half (44 percent) of the landlords had fixed indebtedness averaging approximately $11,700 per plantation, representing over 40 percent of the total value of land, buildings,

Table 18—PLANTATION MORTGAGES, BY SOURCE OF LOAN AND BY RATE OF INTEREST, 1934
(Cotton Plantation Enumeration)

Rate of Interest	Total	Holder		
		Government	Other	Unknown
Total	249	136	52	61
2.5	1	1	–	–
3.0	–	–	–	–
3.5	8	7	–	1
4.0	16	11	3	2
4.5	21	19	–	2
5.0	58	46	–	12
5.5	39	29	2	8
6.0	69	23	19	27
6.5	1	–	–	1
7.0	8	–	5	3
8.0	24	–	20	4
9.0	–	–	–	–
10.0	3	–	2	1
Unknown	1	–	1	–
Average rate	5.6	5.0	6.9	5.8

animals, and machinery. This means that the annual interest charge averaged around $660 (Table 18). In spite of A.A.A. benefits there was a reduction of only $200 in the landlords average fixed indebtedness from 1933 to 1934[1] indicating that the profits of that year were used for paying off old floating debts or for repairs and replacements.

Most of the landlords (72 percent), for whom the holder of the debt was reported, had availed themselves of government facilities for mortgage loans (Table 18).

Trends

In the seven cotton States included in the present survey, the aggregate farm mortgage debt on plantations and small farms

[1]Appendix Table 20 contains data for 1934. Due to the small reduction between 1933 and 1934, data for the former year are not shown.

combined amounted to less than $166,000,000 in 1910 but rose to $637,597,000 in 1928 (Appendix Table 21). Thus the amount of mortgage debt almost quadrupled, representing a 167 percent increase from 1910 to 1920, 33 percent from 1920 to 1925, and over 8 percent from 1925 to 1928. These increases were proportionately greater than in any other area of the United States in the last two periods. The increase from 1910 to 1920 was exceeded only by those in the Mountain and Pacific States.[2]

The ratio of farm mortgage debt to the value of all farms in the United States rose from 9.5 percent in 1910 to 21 percent in 1928. For the seven southeastern cotton States as a whole, the ratio was lower than the United States average in both years, but the difference was much less in 1928 (Appendix Table 22). This increase in the ratio of debt to current value of mortgaged farms was partly due to the great increase in the amount of debt, but was also partly a result of the declining value of land.[3]

In contrast to the rapid rise in farm mortgage debt and in the ratio of debt to value, the frequency of farm mortgages remained relatively constant from 1910 to 1928 throughout the United States. In 1910, 33 percent of all farms in the United States were mortgaged; in 1920, 37 percent; in 1925, 35 percent; and in 1928, 36 percent (Appendix Table 23). However, the proportion of new mortgages incurred on farms not previously encumbered was greater in the South than in other sections of the country. From 1925 to 1928 new mortgages constituted between 5 and 6 percent of all mortgages in the South, whereas in no other area did they constitute more than 3.7 percent.[4]

These figures indicate that the South followed the general trend in the United States by a rapid increase in the volume of mortgage debt since 1910 and an increase in the ratio of debt to the value of all farms. Moreover, although the volume of debt in the seven States under consideration constituted a relatively small proportion of the total farm mortgage debt in the United States, it was increasing at a rate faster than in most other areas from 1910 to 1920, and faster than in any other area from 1920 to 1928. This was particularly evident during the years 1925 to 1928 when the increase in the volume of mortgage debt for the United States had tapered off to 1.2 percent, whereas the increase in the seven southeastern cotton States was 8.4 percent (Appendix Table 21).

The trend of increase in farm mortgage debt was reversed between 1928 and 1930. Total mortgage debt in the United States

---

[2]Wickens, David L., *Farm Mortgage Credit*, U. S. Department of Agriculture, Technical Bulletin 288, February 1932.

[3]Based on reports of the *United States Census of Agriculture* and reports from individual farmers.

[4]Wickens, David L., *op. cit.*

fell to below the 1925 level, from $9,468,526,000 in 1928 to $9,241,390,000 in 1930, a decrease of over $227,000,000 or 2.4 percent. The decrease in the seven southeastern States was proportionately greater, dropping from nearly $638,000,000 to $601,000,000, a decrease of 5.8 percent.[5]

Foreclosures

The precariousness of the mortgagee's situation during the past 15 years of unsettled cotton yield and price is indicated by the fact already mentioned that the land held by corporations, mostly through foreclosures, amounted on an average to 10 percent of the total acreage in farms in the 46 counties sampled (Appendix Table 8). Appendix Table 24 shows the extent of this acreage foreclosed in the sample counties and the type of corporations holding the land in 1934. One-third of the acreage held by corporations was in the hands of insurance companies, about one-fourth in the hands of land banks, a sixth in the hands of depository banks, and the rest in the hands of miscellaneous corporations.

This loss of land through foreclosure is not entirely a depression phenomenon. Studies made in the middle of the 1920-1930 decade indicate that corporation holdings had become an important item by that time.[6]

Some of the factors in foreclosure are indicated in the following summary of a statistical analysis of the farm mortgage activities of nine major lending agencies in five southeastern Alabama counties, including the First Joint Stock Bank, the Federal Land Bank, and seven insurance companies.[7]

1. In general, foreclosures and losses increased as appraised value per acre decreased.
2. In general, there was a negative correlation between the average amount of loan per acre on the various soil types and the percent of loans foreclosed.
3. In all soil types, the land was over-valued by appraisers, the average appraised value being higher than the sale price had ever been. Moreover, since the relation of loan per acre to value per acre was practically the same for all soil types (slightly more than one-third), and since foreclosures were considerably more frequent for land with low appraisal value, the poorest types of soil were over-valued more than the better types.

---

[5] Wickens, David L., *op. cit.*

[6] Raper, A. F., *op. cit.*

[7] Mereness, E. H., *Farm Mortgage Loan Experience in Southeast Alabama*, Agricultural Experiment Station of the Alabama Polytechnic Institute, Bulletin 242, January 1935.

4. In general, there was a direct correlation between the percent of loans foreclosed and the size of farms. Large farms were much poorer risks than small ones, particularly when located on poor soils.
5. In areas such as the one studied, where erosion was a serious factor, the nearer the topography approached a level condition, the better the loan risk.
6. Foreclosures and losses increased consistently as the borrower's equity decreased. This was particularly true on the poorer soils.

LANDLORDS' SHORT TERM CREDIT

Tenants, with the exception of independent renters, usually have their production credit arranged and secured by the landlord. He borrows a sufficient amount to meet the needs of the whole plantation and then carries as a debt against the tenant the amount necessary to finance the tenant's share of production costs and the subsistence advances needed by the tenant family. Croppers and other share tenants, therefore, do not usually have the access to primary sources of credit available to the landlord. Since the growing crop is generally the only security for production credit, the agency furnishing the credit usually holds first lien on the crop and other agencies have no adequate security.

In the case of landlord-tenant agreements, the landlord theoretically provides the credit and holds the lien, both for his share of the net profits and for his credit advances. In the case of landlord-bank or landlord-merchant agreements, the bank or merchant holds first lien against the total crop and looks directly to the landlord for repayment. In short, the landlord obtains production credit direct from the source, and reallocates part of this credit to the tenant, usually at a higher rate of interest.

Rates of Interest

Another peculiarity of the cotton-credit system is that usury laws are inoperative. The legal rate of interest is a fiction. Credit is used, not for 12 months, but for from 3 to 8 months, yet interest is charged at a flat rate as if the loan were used for the full year. The most common practice is to charge a flat 10 percent. That is to say, a loan of $100 will cost $10 regardless of whether the loan runs for 3 or 8 months. If $10 is charged for a $100 loan of 3 months' duration this would be equivalent to $40 for a full year, or 40 percent annual interest.

Both landlord and tenant use money for varying periods. Credit for fertilizer, bought early in the spring, usually extends throughout the crop season. Landlord borrowing to meet payrolls, however, is done from month to month. Credit for his first month's payroll may extend through the entire season whereas he will have money for the last month's disbursements for only 30 or 45 days. Similarly the number of months duration of the tenant's subsistence advances must be averaged to calculate the time his credit runs, since his February advance may not be repaid for 8 months while his August advance may be repaid in 1 month. Obviously, a tenant who receives $10 a month advance from February to August, repaying in September, uses $80 credit but only for an average of 4 months. A charge of $8 interest for this 4 months' use is equivalent to $24 for a full year's use, or 30 percent on an annual basis.

Table 19—LANDLORD CURRENT BORROWING BY NUMBER OF TENANTS, 1934
(Cotton Plantation Enumeration)

Number of Tenant Families	Total Plantations Borrowing[a]		Total Families	Aggregate Borrowing[a]	Borrowing per Plantation	Borrowing per Family
	Number	Percent				
Total	339	100.0	5,033	$782,347	$ 2,308	$155
5 to 10 families	179	53.0	1,115	151,382	846	136
10 to 15 families	61	18.0	716	120,180	1,970	168
15 to 20 families	25	7.0	420	70,293	2,812	167
20 to 25 families	21	6.0	453	56,050	2,669	124
25 to 30 families	16	5.0	436	64,544	4,034	148
30 to 35 families	10	3.0	319	60,000	6,000	188
35 to 40 families	6	2.0	225	31,650	5,275	141
40 families and over	21	6.0	1,349	228,248	10,869	169

[a]Includes all borrowing in addition to government, merchant, fertilizer, and bank loans. See Table 20 and Appendix Table 25.

Of the landlords included in this study, over half borrowed on short term for 1934 production expenses. The average amount borrowed was $2,308, an average of $155 for each family on these plantations (Table 19).[8] The amount borrowed is roughly half of the current operating expense and tenants' subsistence advances.

The banks were predominant sources of landlord credit (Table 20). A few landlords obtained government loans or advances from merchants, and a negligible number purchased fertilizer on credit.[9]

The largest average loans were from government agencies, the next largest from banks, with loans from merchants smallest. Annual interest rates followed the reverse order. The average flat amount of interest was about $100 per year for cost of current borrowing. This, added to the approximately $660 per year as interest on the mortgage debt, resulted in a combined

[8]On a few other plantations tenants financed themselves by direct credit from banks or merchants and a substantial number financed their operations without borrowing.

[9]For data by areas, see Appendix Table 25.

interest charge amounting to almost as much as the landlord's net labor income ($855).[10]

Table 20—LANDLORD BORROWING FOR PRODUCTION CREDIT BY SOURCE OF LOAN,[a] 1934
(Cotton Plantation Enumeration)

Item	Bank Loans	Government Loans	Merchant Loans
Number of plantations borrowing	225	57	48
Total amount of loans	$497,566	$164,214	$91,366
Average duration of loans in months	3.7	3.8	4.6
Average interest rate	15.2	10.4	16.4

[a]Based on Appendix Table 25.

Governmental Lending Agencies

Inasmuch as the Federal government has made special efforts through various lending agencies to alleviate the losses of farmers, analysis of the operation of these agencies in relation to the plantation system is pertinent. It has been noted in this study, as well as in other studies of credit, that loans from public agencies or banks are almost never made to croppers or other share tenants. The reason for this, as has been pointed out, is that the agency supplying the credit must have as security a lien on the tenant's equity in the crop. The landlord already has such a lien for his share of the crop and for reimbursement for his share of the expenses. If the landlord is unwilling to relinquish his claim, the tenant has no security to offer another lending agency. Government agencies, therefore, supply credit only to landlords, who in turn pass the charges along to tenants, usually at a much increased rate of interest.

The Farm Credit Act of 1933 authorized the organization of a production credit system for farmers, consisting of 12 area Production Credit Corporations and numerous local Production Credit Associations. These were organized to make loans to farmers for general agricultural purposes and to rediscount the notes of their borrowers with the Federal Intermediate Credit Banks. The Production Credit Associations are the local lending institutions and 597 of them were in operation as of December 31, 1934 (Appendix Table 26).

In the 7 southeastern cotton States, 147 Production Credit Associations were in operation by the end of 1934: 34 in Georgia, 32 in North Carolina, 30 in Arkansas, 25 in South Carolina, 10 in Mississippi, 8 in Alabama, and 8 in Louisiana. These represented 25 percent of the total number in the United States.

---

[10]See chapter VI.

From the date of organization of such Production Credit Associations through December 31, 1934, a total of 131,621 loans were closed in the United States, amounting to $92,882,000. Of these, 48,301 loans, amounting to $17,137,000, were in the 7 States under survey. These represented nearly 37 percent of the number and over 18 percent of the amount of all such loans closed in the United States during that period.

The average size of Production Credit Association loans was considerably smaller for the southeastern cotton area than for the Nation as a whole, amounting to $355 per loan in contrast to $706 for the United States. Loans averaged only $260 and $267 in North Carolina and South Carolina, respectively, whereas these 2 States reported the greatest number of loans of any States in the country, 11,883 and 10,552, respectively (Appendix Table 26).

Maturities acceptable to the Intermediate Credit Banks through production credit and other credit agencies are arranged to coincide with normal marketing or liquidating seasons, in order to permit notemakers to complete operations in the season for which credit is extended. Ordinarily maturities range from 3 months to 1 year, and under the law may not exceed 3 years.

As of December 31, 1934, the discount rate of Federal Intermediate Credit Banks was 2 percent, and the rate charged borrowers by Production Credit Associations was 5 percent.

In order to continue at the present low rates of interest, it is essential that the Federal Intermediate Credit Banks maintain their present high credit standing among investors, thus enabling them to sell debentures at very low interest rates. This means that loans to Production Credit Associations must be adequately secured and must provide for liquidation at maturity. Consequently these loans are made almost exclusively to farm owners, and then only when ample security is provided.[11]

Congress made funds available for Emergency Crop Production and Seed Loans[12] through Federal appropriations in 8 different years during the period 1921 to 1932. Originally these loans were confined to specified areas affected by disaster, and prior to 1931 the aggregate amount advanced in any 1 year did not exceed $9,000,000.

In 1931, a considerable expansion of the program was begun as a result of the breakdown of commercial banking facilities

---

[11]See *Farm Credit Administration Yearbook, Second Annual Report, 1934*. Appendix Table 43: *Federal Intermediate Credit Banks:* Loans to and discounts for financing institutions, outstanding on December 31, 1934, by States and type of institution. Appendix Table 54: *Production Credit Association's Applications:* Received and submitted to Federal Intermediate Credit Banks, and loans closed from organization through December 31, 1934, and loans outstanding on December 31, 1934, by States.

[12]"Emergency Crop Production and Seed Loans" became "Emergency Crop Production and Feed Loans" in the Act of February 23, 1934.

in rural sections, together with serious crop failures in wide areas, and low prices fo farm commodities. From 1921 to 1930, an aggregate of 71,672 loans, amounting to $8,909,000, were made in the 7 southeastern cotton States, constituting 59 percent of both the number and value of all such loans made in the United States. In 1931, 197,117 loans were made, amounting to $25,174,000; in 1932, 250,899 loans, amounting to $25,328,000; in 1933, 365,209 loans, amounting to $30,711,000; and in 1934, 214,132 loans, amounting to $13,922,000. These loans constituted 51 percent of the number and 44 percent of the value of all such loans made in the United States from 1931 through 1934 (Appendix Table 27).

It is significant that 579,341 loans were made, aggregating $44,633,000, in the Eastern Cotton Area alone during the first 2 years of the A.A.A. cotton reduction program. The concentration of such loans in the southeastern cotton States is indicated by the county data in Figure 15. On a state basis, the ratio of loans to farms ranged from 4 percent in Alabama in 1930 to 38 percent in South Carolina in 1933 (Appendix Table 28).

An indication of improved financial conditions in the South may be obtained from Appendix Table 29, showing the proportion of loans collected during the years 1931 to 1934. In the seven States, 46 percent of the loans of 1931, 54 percent of those of 1932, and 80 percent of those of 1933 were collected by November 30, 1933. Eighty-eight percent of the 1934 loans were collected by December 31 of that year. These seven States made a considerably better showing, with respect to the proportion of loaned funds collected, than did the rest of the United States. In 1931, when loans in these seven States constituted 45 percent of the value of all loans in the United States, collections amounted to more than 46 percent of the total amount collected. In 1932, when those States obtained 39 percent of the value of all loans, collections amounted to 49 percent of the total collected. In 1933, when loans represented 54 percent of the United States total, collections amounted to 72 percent of total collections; and in 1934, when loans in these States included but 37 percent of the value of all loans, collections amounted to 69 percent of the total amount collected (Appendix Tables 27 and 29).

Although the proportion of cash collections to total collections in the South was consistently greater than the proportion of all loans made in these States from 1931 to 1934, the difference was considerably more marked in 1933 than in the 2 preceding years, and very much more marked in 1934 than in 1933. This not only points to improved financial conditions in cotton areas during these years, but indicates that a considerable

FIG. 15– AMOUNT OF EMERGENCY CROP AND FEED LOANS, BY COUNTIES

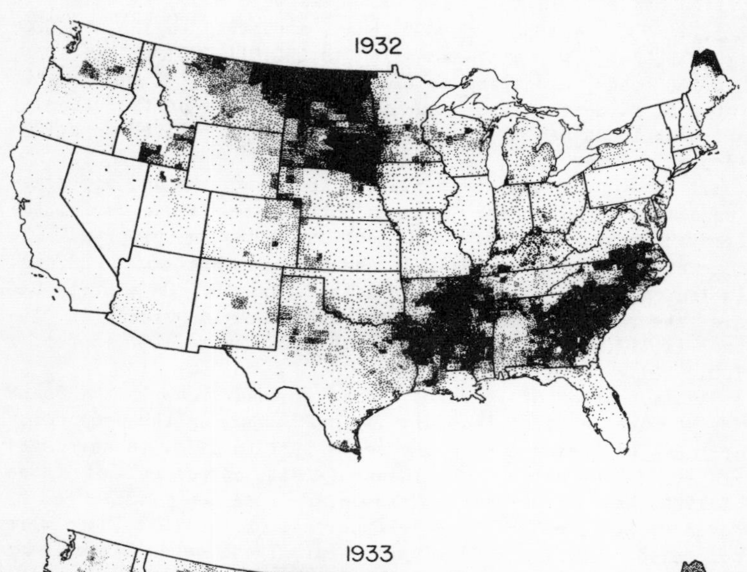

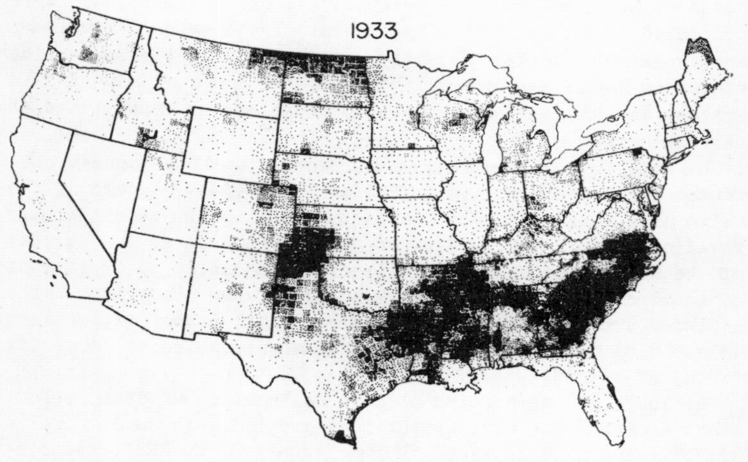

Each dot represents $ 1,000 or fraction thereof

Farm Credit Administration

Division of Finance and Research, No. 1387

amount of the A.A.A. cotton reduction and benefit payments was used to pay current and past debts.[13]

TENANTS' SHORT TERM CREDIT

Since there is no landlord lien against their crop, independent renters can negotiate loans wherever they can obtain credit, but croppers and other share tenants are practically forced to secure credit through the landlord. They not only need credit

Table 21—SUBSISTENCE ADVANCE PRACTICE OF PLANTATIONS, BY AREAS, 1934
(Cotton Plantation Enumeration)

Area	Total Plantations	Number of Plantations Making Advances	Advance per Family per Month	Plantations Reporting Number of Months Advanced[c]	Months Advanced							Average Number of Months Advanced
					Less Than 5	5	6	7	8	9 or 10	11 or 12	
Total	646	553	$12.80	530	20	81	206	49	70	76	28	6.9
Atlantic Coast Plain	56	44	13.70	44	-	2	3	3	10	20	6	9.0
Upper Piedmont	40	37	10.10	31	2	7	7	1	4	9	1	6.8
Black Belt (A)[a]	112	98	8.20	93	4	9	29	14	10	16	11	7.5
Black Belt (B)[b]	99	80	9.40	79	-	15	29	13	12	8	2	6.8
Upper Delta	133	122	17.30	114	3	7	77	8	4	12	3	6.6
Lower Delta	50	34	9.70	34	8	6	9	-	8	3	-	6.1
Muscle Shoals	22	17	17.20	16	-	9	-	1	1	5	-	6.8
Interior Plain	30	29	17.30	28	1	4	9	1	11	-	2	7.0
Mississippi Bluffs	47	46	12.60	45	1	6	30	4	3	1	-	6.1
Red River	28	20	16.80	20	-	4	5	2	6	1	2	7.3
Arkansas River	29	26	15.30	26	1	12	8	2	1	1	1	5.9

[a] Cropper and other share tenant majority.
[b] Renter majority
[c] Data on actual number of months tenants were advanced not available for 23 of the 553 plantations reporting an advance.

for their seeds, feed, and fertilizer, but most of them have exhausted the income from their past crop by February or March and must be advanced supplies for subsistence until the next crop is harvested.

Table 21 summarizes the subsistence advance practice of plantations. Of those surveyed 86 percent made advances to tenants, averaging $12.80 per family per month for an average of 7 months. Very few plantations carried their tenants less than 5 months and only about one-fifth of the plantations made advances for more than 8 months.

As has been pointed out, this current obligation is not considered by the tenant as a debt, but as income drawn in advance, and if the year's operation is successful the transaction is

---

[13] See *Farm Credit Administration, First Annual Report, 1933*, pp. 51-52. Appendix Table 35: *Emergency Crop Production and Seed Loans:* Number and amount of loans from 1921 through 1932 and cash collections through November 30, 1933, by States; and *Farm Credit Administration, Second Annual Report, 1934*, pp. 63-67. Appendix Table 64: *Emergency Crop and Feed Loans:* Loans made in each State during 1933 and 1934 and balances of these loans outstanding on December 31, 1934.

balanced off when the crop is marketed. However, if the proceeds of the crop are not sufficient to pay off advances, the accumulating obligation is considered as a debt.

Tenant Debts

Even in a relatively good year a considerable proportion of tenants fail to repay their advances. In the present study tenants were asked whether they lost, broke even, or gained during

Table 22—TOTAL DEBT[a] OF PLANTATION FAMILIES, BY TENURE, 1930–1934
(Cotton Plantation Enumeration)

Tenure and Year	Total Families[b]	Total Families in Debt	Percent in Debt	Families Reporting Amount of Debt	Total Debt	Debt per Family Reporting Amount of Debt
All tenures						
1934	4,156	603	14.5	554	$49,127	$ 89
1933	3,274	470	14.4	407	46,168	113
1932	2,846	423	14.9	323	35,985	111
1931	2,824	362	12.8	246	29,655	121
1930	2,704	350	12.9	262	37,459	143
Croppers						
1934	2,812	280	10.0	265	14,447	55
1933	2,154	246	11.4	222	16,386	74
1932	1,997	235	11.8	187	14,842	79
1931	1,836	215	11.7	152	13,070	86
1930	1,743	234	13.4	169	18,546	110
Other share tenants						
1934	701	121	17.3	99	11,854	120
1933	513	91	17.7	69	11,767	171
1932	273	77	28.2	48	8,617	180
1931	425	71	16.7	39	7,310	187
1930	409	45	11.0	45	10,428	232
Renters						
1934	643	202	31.4	190	22,826	120
1933	607	133	21.9	116	18,015	155
1932	576	111	19.3	88	12,526	142
1931	563	76	13.5	55	9,275	169
1930	552	71	12.9	48	8,485	177

[a]Debt at end of specified year, including debt of past years.
[b]Families for which data were available.

the past 5 years. This inquiry revealed that 12.7 percent of those reporting[14] came out in debt at "settling time" in 1930 and that this percentage had declined to 7.3 in 1933. Among croppers 14.0 percent reported debts at "settling time" in 1930 in comparison with 6.5 percent in 1933 (Appendix Table 30). The number of tenants in debt is further indexed by Table 22 which indicates that 14.5 percent were in debt at the end of the 1934 crop season. The lowest proportion of indebtedness for any tenure group in 1934 was among croppers (10 percent). Other share tenants followed with 17.3 percent in debt while 31.4 percent of the renters were in debt.

[14]Including wage hands.

The percentage of tenants in debt is materially reduced by migration. When a cropper moves to a new location he usually considers that he is starting with a clean slate and the landlord writes off old debts unless he is able to recover by levying on the tenant's personal property.

The average amount of indebtedness per tenant in debt in 1934 was $89 (Table 22). The debts of tenants owning work stock (share tenants and renters) were much higher than those of croppers--$120 and $55, respectively. This is explained by the fact that the cropper's expenditures are supervised, the more efficient crop operation of the cropper group, and the larger borrowing of tenants owning work stock. For all classes of tenants there was a substantial reduction in the average debt from 1930 to 1934, due probably in the earlier years to the increased supervision of subsistence advances as credit conditions became progressively more constricted, and in 1933 and 1934 to the application of increased profits to payment of back debts. The cropper average debt was cut in half from 1930 to 1934, the greatest reduction being between 1933 and 1934.

Tenant Interest Rates

The importance of credit operations in landlord-tenant relationships warrants going beyond the first-hand data gathered in this study. Landlord advances are made to tenants in three ways or by a combination of these: direct cash advances, advances in the form of merchandise, or establishment of a credit account at a merchant store where tenants are permitted to purchase supplies up to a stipulated amount per month.

In a study of 588 croppers on 112 North Carolina farms in 1928[15] it was found that no cropper borrowed a single dollar directly from banks or public lending agencies. All croppers were furnished farm supplies, including fertilizer, either by landlord or merchant.[16] Eighty-two percent of the croppers received cash advances from the farm owner, averaging $109 each at a cost of twenty-one percent in interest. Sixty percent of the croppers received household supplies direct from the landlord at an average value of $113 per cropper, costing fifty-three percent in interest. Forty-one percent of the croppers received household supplies from merchants on the landlord's guarantee, averaging $54 per cropper, and costing an average of seventy-one percent in interest (Appendix Table 31).

Thus, in this North Carolina study, subsistence advances to croppers were most commonly in the form of cash from the landlord.

---

[15] Wooten, H. H., *Credit Problems of North Carolina Cropper Farmers*, North Carolina Agricultural Experiment Station, Bulletin 271, May 1930.

[16] No data are available as to the proportion from each source.

Credit in the form of merchandise from the landlord was of secondary importance and a still smaller proportion of croppers used merchant credit on the landlord's guarantee. Whereas the cost of every type of credit was extremely high, the cash and merchandise advances from the landlord cost less than did direct merchant credit. Cash advances were the least expensive and were usually given at a 10 percent flat rate in time charges, while goods advanced commonly carried a flat 20 or 25 percent in time charges.

The duration of the loan accounts for these high per annum interest rates and for the differences between types of credit advanced. Cash advances by the landlord averaged 4.9 months; farm supplies by owner and merchant averaged 8.3 months; household supplies by farm owners averaged 4.8 months, and by merchants on the owner's guarantee 4.7 months. The average duration of all advances to croppers was slightly less than 6 months (Appendix Table 31).

Amounts advanced to the croppers during the year aggregated more than 63 percent of their cash farm income, and interest paid on those advances accounted for more than 10 percent of that income. The cash farm income was given at $766 per cropper.

The North Carolina study showed that every type of credit cost croppers more than it did owners. For cash loans from banks, owners paid an average of 6.5 percent interest, while croppers paid 21 percent for cash loans from individual owners at periodic intervals throughout the season.[17] For farm supplies including fertilizer, owners paid at the high average rate of 30 percent per annum, but croppers paid 32 percent. In addition, croppers paid interest on subsistence advances amounting to over 50 percent to landlords and over 70 percent to merchants.

The differences in cost of credit to tenants and owners are generally explained as due to the fact that tenants involve considerably more risk. High rates are charged to allow for defaults, good risks among tenants compensating for the bad.

As a matter of fact, according to findings of the various credit studies, interest rates are placed so high that the amount of actual default, when subtracted from total interest payments, still results in excessive interest profits except in years of crop disaster. The average flat rate per dollar charged the 588 North Carolina croppers was 19 percent, and the average loss to landlords and merchants on total advances was but 5 percent. This left a flat rate per dollar net gain of 14 percent. Since the average duration of all loans to croppers was slightly under 6 months, the net gain to owners and merchants

---

[17] Wooten, H. H., *op. cit.*, p. 18.

on cropper credit extended was at the rate of approximately 28 percent per annum.

It is evident, then, that landlords and merchants are taking care to keep the interest rate well above any possibility of loss from defaulting tenants. The tenant who makes good his loans pays a high charge to allow his defaulting neighbor to slip by. Moreover, the sooner a loan is paid after it becomes due, the higher is the per annum rate of interest which the tenant pays. Since landlord and merchant credit continue to be relatively easy to obtain, there is no decided incentive for prompt payment of debts. Moreover, the hard terms and the knowledge gained by experience that the store bill will eventually

Table 23—PLANTATIONS MAKING SUBSISTENCE ADVANCES: AMOUNT, DURATION, AND ANNUAL RATE OF INTEREST, BY AREAS, 1934
(Cotton Plantation Enumeration)

Area	Total Plantations Reporting[a]	Amount of Advances	Average Months Duration	Annual Rate of Interest
Total	535	$634,980	3.6	37.1
Atlantic Coast Plain	44	29,924	4.3	19.0
Upper Piedmont	33	8,436	3.4	18.7
Black Belt (A)[b]	88	39,424	3.8	19.5
Black Belt (B)[c]	81	22,449	3.4	22.9
Upper Delta	119	280,274	3.3	40.6
Lower Delta	34	25,031	3.0	44.8
Muscle Shoals	16	8,147	3.4	14.9
Interior Plain	29	42,818	3.6	36.3
Mississippi Bluffs	45	46,061	3.1	40.3
Red River	20	43,154	3.7	30.0
Arkansas River	26	89,262	2.9	55.0

[a]Thirty-five additional plantations made advances but data on amounts advanced were not available.
[b]Cropper and other share tenant majority.
[c]Renter majority.

take a large proportion of the crop anyway, often cause the tenant to become discontented, to produce poor crops, and to be indifferent toward repayment.

In short, high rates of interest involved in the system of landlord furnishing and merchant credit obtained on the landlord's guarantee constitute major factors preventing a rise of tenants on the agricultural ladder. The fact that the credit system is based on cash crop liens is also a stumbling block to diversification. The system, furthermore, shifts the power to market the crop from tenant to landlord and the landlord himself is often forced to sell, regardless of the condition of the market, in order to meet his maturing obligations.

First-hand information on tenant interest rates was obtained in this study only on the interest charged by landlords on subsistence advances. Flat interest rates of the tenant's share of expenses are usually similar to interest on subsistence advances but annual rates are lower, as fertilizer loans, for example, run throughout the crop season. The weighted average annual interest rate paid by all tenants on subsistence advances in 1934 was 37.1 percent (Table 23).[18] This is in contrast with the average rates ranging from 10 percent on government loans to 16 percent on merchant loans paid by landlords (Appendix Table 25). There was considerable local variation in interest rates—in the eastern part of the area surveyed they ranged from 15 to 23 percent and in the western part they were around 40 percent.

---

[18]This is a flat rate of $11.16 for an average duration of 3.6 months, *i.e.*, for each $100 advanced the tenant pays $11.16 but uses the credit for an average of only 3.6 months.

Chapter VI

INCOME

The complex internal organization of plantations, together with variations among individual units, presents a difficult problem to persons interested in measuring the earning capacity of the plantation system.

As indicated in preceding chapters, a plantation may be a relatively small unit operated by a resident landlord and four or five tenant families, all with about the same rental contract, or it may be a unit of thousands of acres farmed by croppers, other share tenants, cash or standing renters, and wage hands, all working under the supervision of hired "riding bosses", employed by an absentee landlord or corporation owner. There may be a commissary or store on the plantation operated primarily for the purpose of selling goods to the families on the plantation, but this store may also sell to persons in no way connected with the plantation. The plantation may have a cotton gin for ginning cotton produced on the plantation, but it may do ginning for others as well.

To understand the factors in landlord and tenant income it is necessary to unravel these complexities in the plantation organization, and to find records of financial transactions. Many plantation operators do not keep permanent financial records of any kind, and in instances where written records are kept they are often not adaptable to the statistical procedures involved in studying a large number of cases. Croppers, tenants, and laborers on the plantations usually have no written records whatsoever, and often are able to give only very rough estimates of their past financial transactions.

For this survey, in order to simplify the analysis and to have results as nearly homogeneous as possible, the financial records obtained were limited solely to farming operations for the crop year 1934. The financial results of non-farming enterprises such as commissaries or cotton gins were not included, and no inventory was taken of livestock, farm products, or equipment. Neither was any attempt made to evaluate perquisites such as free house rent and fuel, commonly furnished plantation families. In other words, the financial data presented in this chapter pertain almost wholly to current cash receipts from crops and livestock and to current cash expenditures, together with a statement of the value of commodities produced for home consumption.

Although this procedure does not give a picture of all items of income and expenditure, it nevertheless includes the major items and makes possible comparisons between areas and types of plantations. It gives a reasonably complete view of the current financial results of the farming operations on the plantations studied, since minor items of income which are omitted are balanced by minor items of expenditure omitted.

Table 24—GROSS INCOME OF 645 PLANTATIONS,[a] BY SOURCE, 1934
(Cotton Plantation Enumeration)

Source of Income	Amount		
	Total	Per Plantation	Percent of Total
Total	$6,126,570	$9,498	100.0
Crop sales			
Shared by landlord and tenants[b]	4,310,256	6,683	70.4
Unshared by landlord and tenants[b]	1,816,314	2,815	29.6
Landlord	953,539	1,478	15.6
Sale of livestock products	158,451	246	2.6
Cash rent from land	97,569	151	1.6
A.A.A. payments	530,279	822	8.7
Home use	167,240	259	2.7
Tenants[b]	862,775	1,337	14.1
Sale of crops and livestock products	42,160	65	0.7
A.A.A. payments	69,756	108	1.1
Home use	750,859	1,164	12.3

[a]Data not available for one plantation.
[b]Unless otherwise specified the term "tenant" includes both share-croppers and other share tenants.

The earning capacity of the plantation system will be discussed in this chapter from two standpoints, that of the plantation as a complete unit and that of the families involved, including the landlord, and the cropper, tenant, and laborer families under his supervision.

PLANTATION INCOME

This section, pertaining to "plantation income", is concerned with: (1) the current farming receipts of the plantation from commodities produced for home consumption and from sales of goods, mainly off the plantation; and (2) current farming expenses paid by all persons on the plantation, mainly to persons or agencies outside the plantation. In some cases it was impossible to separate minor items of receipt or expense involving persons on the plantation from those involving outside persons or agencies. However, all major intra-plantation transactions were eliminated.

Gross Income

Plantations are, in the main, highly commercialized organizations depending for their income primarily upon the sale of crops. This is evident from the fact that approximately 70 percent of the total 1934 gross income of the 645 plantations for which complete records were obtained was from crop sales, in which both the landlord and tenant shared (Table 24).

The largest single item of income unshared by landlord and tenants[1] was received from the Agricultural Adjustment Administration for compliance with its acreage reduction program. From these payments the landlord received an average of $822 per plantation, compared with $108 per plantation received by all tenants together. These combined amounted to about 10 percent of the 1934 plantation income.

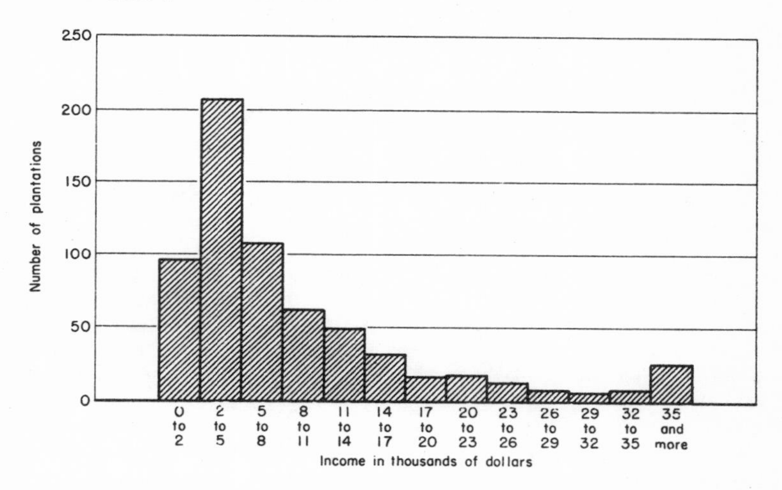

FIG. 16 - TOTAL GROSS INCOME OF 645 COTTON PLANTATIONS 1934

Source Appendix table 32 AF-1457, W.P.A.

The value of the products produced for home use on the plantation amounted to 15 percent of the total, of which 2.7 percent was landlord production and 12.3 percent was tenant production. The total gross income averaged about $9,500 per plantation.[2]

This average gross income per plantation is deceptively high because of the inclusion of a few high income plantations in the study. The plantations were concentrated in the low income groups (Appendix Table 32 and Figure 16), but a substantial

---

[1]Exclusive of home use production.

[2]The rent received from cash renters was included in the gross income figures, but in all other respects the cash renter was not considered an integral part of the plantation organization.

number of units had a gross income of $20,000 or more. The income of the median (or middle) plantation was $5,540. Approximately 64 percent of the plantations had a gross income in 1934 of less than $8,000, almost 50 percent had an income of less than $5,000, and 15 percent had an income of less than $2,000. On the other hand, the gross income of approximately 12 percent of the plantations was more than $20,000 per unit.

Variations in total gross income among plantations are related to a large extent to the geographical distribution of the plantations surveyed. The average gross income per unit ranged from a high of $26,963 in the Arkansas River area to a low of $3,732 in that part of the Black Belt area of Alabama, Georgia, and Mississippi where cash renters are in the majority among tenants (Table 25). The average gross income in 7 of the 11 areas surveyed was lower than the average for all areas ($9,498). The four areas in which the gross income per plantation was higher than the average for all areas were the Arkansas River area, the Upper Delta area, the Red River area, and the Interior Plain area. The 29 plantations in the Arkansas River area had a total gross income per plantation almost 3 times as large as the average of all areas.

The fact that size of plantation is the chief determining factor of gross income is indicated by the fact that the per acre income (equating size) does not show the wide fluctuations shown by the total gross income. The data in Table 26 contrast for each area the 25 percent of the plantations having the highest gross incomes per plantation with the 25 percent having the lowest. Although there was a great difference in the average income per plantation between the two groups of plantations, the difference in gross income per capita between these two groups was not extreme. The average gross income per capita for the one-fourth of the plantations in each area with the highest gross income per unit was $196, which is only $23 more than the average for all plantations, and $65 more per capita than for that group of plantations with the lowest gross income per plantation.

Net Income

If the current cash expenses, paid by both landlords and tenants for making the 1934 crop, are subtracted from the total gross income, the net income figures shown in Table 25 are obtained. The expense items which have been subtracted from the total gross income include expenditures for feed and fertilizer, interest on short term loans, wages of miscellaneous hired workers, cost of ginning, and other similar cash outlays, which were directly incurred by either landlord or tenant in producing the 1934 crop. Expenses such as interest on long term farm mortgage

Table 25—PLANTATION GROSS AND NET INCOME, BY AREAS, 1934
(Cotton Plantation Enumeration)

Area	Total Plantations Reporting[a]	Number of Persons per Plantation	Acres per Plantation		Plantation Gross Income				Plantation Net Income			
			Total	Crop	Total	Per Plantation	Per Capita	Per Crop Acre	Total	Per Plantation	Per Capita	Per Crop Acre
Total	645	55	905	385	$6,126,570	$ 9,498	$173	$24.74	$3,885,742	$ 6,024	$110	$15.69
Atlantic Coast Plain	56	42	785	294	464,173	8,289	197	27.91	299,229	5,343	127	17.99
Upper Piedmont	40	33	437	211	193,384	4,835	147	22.91	142,124	3,553	108	16.84
Black Belt (A)[b]	112	34	785	275	541,699	4,837	142	17.59	345,708	3,087	91	11.23
Black Belt (B)[c]	99	27	840	256	369,484	3,732	138	15.17	238,668	2,411	89	9.80
Upper Delta	133	92	1,031	563	2,278,533	17,132	186	30.43	1,408,919	10,593	115	18.82
Lower Delta	49	35	1,117	207	205,969	4,203	120	20.01	157,387	3,212	92	15.30
Muscle Shoals	22	25	555	225	91,948	4,179	167	18.57	69,861	3,176	127	14.12
Interior Plain	30	68	1,160	523	335,528	11,184	164	21.38	238,229	7,941	117	15.18
Mississippi Bluffs	47	57	786	378	418,949	8,914	156	23.58	265,657	5,652	99	14.95
Red River	28	77	901	531	444,965	15,892	206	29.93	243,431	8,694	113	16.37
Arkansas River	29	143	1,722	998	781,938	26,963	189	27.02	476,529	16,432	115	16.46

[a] Data not available for one plantation.
[b] Cropper and other share tenant majority.
[c] Renter majority.

loans, or costs chargeable to deterioration or improvements were not included.

The average net income per plantation, according to these calculations, was $6,024. The variation among the different areas covered in the survey was similar to the distribution of the gross income per plantation (Figure 17). Most of the plantation areas with the highest gross incomes per plantation unit also had the highest net incomes per plantation unit, and similarly those with the lowest gross incomes also had the lowest net incomes.[3] About the only significant exception was in the case of the cropper-majority Black Belt area, which ranked seventh in average gross income per plantation and tenth in average

Table 26—GROSS INCOME FOR THE ONE-FOURTH OF THE PLANTATIONS IN EACH AREA WITH THE HIGHEST AND THE LOWEST GROSS INCOME PER PLANTATION, 1934
(Cotton Plantation Enumeration)

Area	Total Plantations in Each Income Group	Gross Income for One-fourth of Plantations in Each Area with Highest Gross Income per Plantation			Gross Income for One-fourth of Plantations in Each Area with Lowest Gross Income per Plantation		
		Per Plantation	Per Capita	Per Crop Acre	Per Plantation	Per Capita	Per Crop Acre
Total	162	$21,143	$196	$29.69	$2,728	$131	$14.50
Atlantic Coast Plain	14	15,478	248	43.78	2,759	111	11.40
Upper Piedmont	10	9,116	188	29.98	2,365	92	14.97
Black Belt (A)[a]	28	9,641	168	22.08	1,608	101	9.17
Black Belt (B)[b]	25	8,353	158	25.04	924	151	4.75
Upper Delta	33	39,071	208	34.19	4,653	135	23.35
Lower Delta	12	12,119	151	28.40	575	113	4.76
Muscle Shoals	6	7,076	186	19.44	1,894	167	12.47
Interior Plain	8	26,393	186	22.08	2,737	100	18.27
Mississippi Bluffs	12	22,875	199	26.97	1,888	94	14.97
Red River	7	36,795	204	34.00	3,644	216	18.58
Arkansas River	7	59,725	197	28.90	9,941	211	25.67

[a]Cropper and other share tenant majority.
[b]Renter majority.

net income per plantation (Table 25). The difference is accounted for largely by the greater outlays for fertilizer and the lower yields of this area as compared with other areas.

Income per Capita

A distribution of the total plantation income, such as is involved in computing the income on a per capita basis, does not show the actual income of the persons on the plantations studied because the landlord shares in the income of all the tenants. Nevertheless, the per capita income figures are significant indices of the earning capacity of the plantation in relation to the persons directly dependent upon it. When the gross plantation

---

[3]It was not true in every instance that the one-quarter of the plantations with the highest or the lowest gross incomes were identical with the one-quarter having the highest or lowest net incomes, respectively. Consequently, the plantations referred to in Table 26 are not identical with those included in Appendix Table 33.

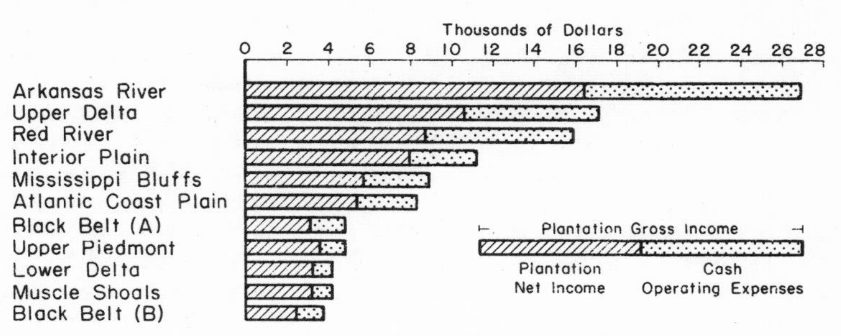

FIG. 17-AVERAGE GROSS AND NET PLANTATION INCOME, BY AREAS, 1934

Source: Table 25 AF-1489, W.P.A.

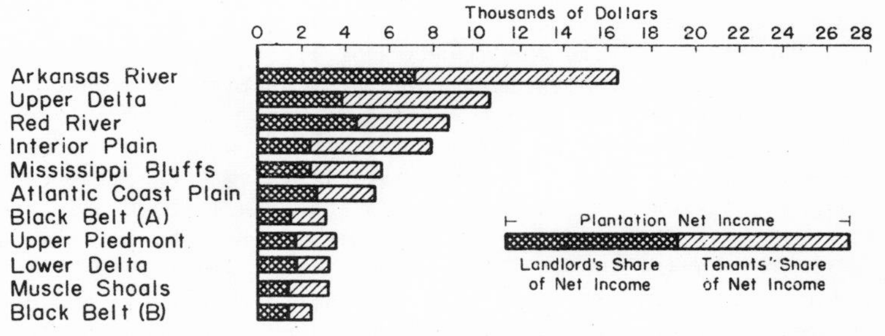

FIG. 18-AVERAGE NET PLANTATION INCOME AND LANDLORD NET INCOME, BY AREAS, 1934

Source: Tables 25 and 28 AF-1491, W.P.A.

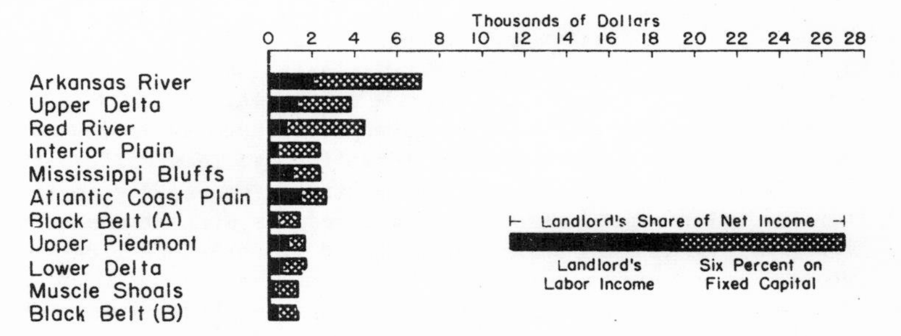

FIG. 19-AVERAGE LANDLORD NET INCOME AND LABOR INCOME, BY AREAS, 1934

Source: Appendix table 36 AF-1493, W.P.A.

income is put on a per capita basis, the average is $173 per person, and the range is from $120 in the Lower Delta to $206 in the Red River area (Table 25). It appears from these figures that if the total gross plantation income had been divided equally among all persons on the plantations, each person would have received an equivalent of about 47 cents per day. In the area with the highest plantation income per capita, such a division would have given each person approximately 56 cents per day, and in the Lower Delta the figure would have fallen to 33 cents.

The average plantation net income per capita was $110 for all areas, and ranged from a high of $127 in the Atlantic Coast Plain and Muscle Shoals areas to a low of $89 in the renter majority Black Belt area. This was the income which the plantation population had available for paying the annual increment of the fixed charges such as long term mortgage interest, repairs, and similar costs, and for living expenses. These figures reflect the low productivity of the small producer units which make up the plantation, and indicate the seriousness of the problem of raising the standard of living among the tenant and farm laborer classes in the southern plantation areas.

In this connection it is pertinent to point out that a larger net income per plantation does not necessarily mean a correspondingly larger average income per capita. Although the relatively large plantations had larger incomes than the average and consequently, slightly higher incomes per capita, the difference in the per capita income was by no means so great between the two groups of plantations as was the difference in income per plantation or per crop acre. The income per crop acre was approximately twice as large on the one-fourth of the plantations with the highest gross income as on the one-fourth with the lowest gross income; the income per person, however, was only about 50 percent larger.

This situation reflects to a great extent the low productivity of the one-family tenant and cropper farms which are the principal component parts of the total plantation unit. Even on the large, well-managed, and efficient plantations with high incomes, the amount of labor required for producing cotton and tobacco is almost as great as on the smaller and less efficiently operated plantations in the same area. Hence, due to the increase in the number of persons employed, the plantations with higher incomes do not show a corresponding increase in per capita income.[4]

It is obvious from the data that even the plantations with the highest net income per unit (Appendix Table 33) afford a

---

[4]The 25 percent of the plantations with the largest incomes had approximately five times as many persons on them on the average as did the 25 percent with the lowest incomes.

low standard of living for the persons on those plantations, and that among the poorest or most inefficiently operated units, the standard of living supported by the plantation system is extremely low compared to that among farmers in other areas of the country. In the Lower Delta area, for instance, there were 12 plantations among those surveyed which had a net income for the year, above current cash operating expenses, of only $46 per person. When the net income above current farming expenses for a family of five persons averages no more than $230 per year, it is evident that the standard of living of the tenants and laborers on these plantations is far below what is generally recognized as acceptable.

Factors Related to Plantation Income

In addition to the size of the plantation,[5] plantation income appears to be related to the crop acres per plantation, the total cotton acres per plantation, the proportion of crop land in cotton, the productivity of the land, and managerial efficiency.

In analyzing the data for the plantations within each area, it was found that the larger than average plantations had a larger than average gross income per plantation in every area. The gross income per plantation increased in relation to an increase in the acreage per plantation at different rates in different areas. Moreover, there were some individual plantation units relatively small in size that had relatively high gross incomes. However, these exceptions were not numerous and on the whole the large plantations had relatively high gross incomes.

It was also found that the plantations with the largest amount of land in crops usually had the highest gross income per plantation. This would be expected inasmuch as gross income is made up primarily of receipts from the sale of crops. In many farm management surveys of individual family-sized farms where receipts from crop sales were the main source of income, the same relationship has been found.

The largest plantations and those with the largest amount of land in crops also had the highest net incomes. In other words, a high net income is usually associated with a high gross income.

An illustration of the manner in which the size of the plantation is related to the net income is given in Appendix Table 34. The one-fourth of the plantations in each area which had the highest net income above current cash operating expenses also had: (1) a greater number of total acres per plantation in every area except two; (2) larger acreages in crops in every

---

[5]See p. 68.

area; and (3) a larger number of persons per plantation in every area, as compared with the one-fourth of the plantations with the lowest net income. Among the plantations with the lowest net income per plantation the total acreage per unit was smaller than the average in all areas except two; the crop acreage per plantation was less than the average in every area except one; and the number of persons per plantation was smaller in every area (Table 25 and Appendix Table 34).

The amount and proportion of land devoted to the production of cotton were also found to be closely related to the plantation income per unit. All plantations covered by the survey produced some cotton, and a large percentage of them were dependent almost entirely upon cotton production. The year 1934 was, on the whole, a prosperous year for cotton producers (Appendix Table 17). The yield was not far from average, and the price of cotton, relative to the goods purchased for use in the production of cotton, was favorable. Hence, it is not surprising to find among the plantations studied in 1934 that those units with higher than average income also had a higher than average acreage of cotton per unit. A detailed analysis of the individual plantations in this survey indicated that there was a much closer relationship between gross or net income per plantation and cotton acreage than there was between income and total acres or total crop acres.

Plantations with a large acreage in cotton, and this usually means a large percentage of crop land in cotton, were the units making the largest income in 1934 (Appendix Table 35). The one-fourth of the plantations in each area with the highest net income per unit also had: (1) more cotton acres per plantation than the average in all areas except one; (2) a higher than average percentage of crop land in cotton in all areas except two; and (3) more cotton acres per person on the plantation than the average in all areas except three. Conversely the one-fourth of the plantations with the lowest net income per plantation had, with a few exceptions, a lower acreage in cotton, a lower percentage of crop land in cotton, and a lower acreage of cotton per person than the average for all plantations.

Unfortunately, data are not available for the years when the production of cotton was not so favorable as it was in 1934. Hence, there is no way to check the general rule for periods when the price of cotton was low in relation to the goods and services purchased by cotton farmers. It appears not unlikely, however, that the units most highly specializing in cotton production have the highest net income of all units during years when cotton prices are not out of line with other prices, and have the lowest net income of all units during depression years when cotton prices are relatively low.

The influence of land productivity on plantation income is indicated in Tables 25 and 26. By the fact that plantations with high total gross incomes also tend to have high per acre gross incomes, it is evident that increased productivity as well as increased size is measurably related to plantation income.

OPERATOR'S INCOME

As was pointed out above, the gross plantation income was made up of the value of products used by landlord and tenants for home consumption plus the total cash receipts of the landlord and tenants from the sale of crops and livestock, from the renting of land by the operator to cash renters (who were not considered an integral part of the plantation unit for this phase of the study), and from rental and benefit payments received from the A.A.A. This method included the money value of the principal goods produced on the plantation and sold, mainly outside the plantation, or used for home consumption. Similarly, the plantation net income was arrived at by deducting cash outlays to persons or agencies outside the plantation organization, and expenditures for operating purposes made by both landlord and tenants, except interest on mortgages.

This procedure of studying plantation income is similar to that ordinarily followed with respect to family-sized farms. It does not, however, take account of transactions between persons on a given plantation. For instance, it is common practice for the operator to extend cash or commodity credit in the nature of subsistence advances to the tenants on his plantation, for which he charges interest. Such interest is properly an item of cash income to the operator and of cash expense to the tenants. The transaction is solely between persons on a given plantation. Although such transactions were of minor importance in the total, the incomes of individual operators or tenants were significantly affected by them. Hence, the present discussion of the operator's income will take account of such transactions. The major item not heretofore included in the plantation income calculations, which will now be included among the operator's receipts, is that of interest on subsistence advances made by the operator to tenants and croppers.

Gross Cash Income

The shared portion of the operator's income is arrived at in the following way. During the year the operator charges to each tenant: (1) the amount of subsistence advanced, (2) the tenant's share of production expenses, one-half in the case of croppers and two-thirds in the case of other share tenants, and (3) interest on these advances. At the end of the year the operator sells the crop of each tenant, deducts his portion of the proceeds, and also deducts from the tenant's portion the charges specified above. In case acreage is cultivated by wage hands, the entire proceeds are retained by the landlord. In case of land rented, only the stipulated amount of rent can be considered as landlord income.

The average gross cash income of the operators of plantations covered by the survey amounted to $5,095[6] per plantation (Table 27). The lowest average gross cash income received by the operators was in the Muscle Shoals area, and the highest was in the

Table 27—OPERATOR'S GROSS CASH INCOME, BY AREAS, 1934
(Cotton Plantation Enumeration)

Area	Total Plantations Reporting[a]	Operator's Gross Cash Income	
		Total	Average
Total	645	$3,286,485	$ 5,095
Atlantic Coast Plain	56	273,624	4,886
Upper Piedmont	40	98,751	2,469
Black Belt (A)[b]	112	286,750	2,560
Black Belt (B)[c]	99	224,073	2,263
Upper Delta	133	1,131,822	8,510
Lower Delta	49	118,497	2,418
Muscle Shoals	22	43,436	1,974
Interior Plain	30	148,067	4,936
Mississippi Bluffs	47	203,958	4,340
Red River	28	293,324	10,476
Arkansas River	29	464,183	16,006

[a]Data not available for one plantation.
[b]Cropper and other share tenant majority.
[c]Renter majority.

Arkansas River area; in 8 of the 11 areas the average operator's gross cash income was lower than the average for all areas. The operator's average gross cash income in the Upper Delta, Red River, and Arkansas River areas was far above that in the other areas studied.

For the entire group of plantations, it was found that the gross cash income of the operator was less than $2,000 on 38 percent of the units; between $2,000 and $5,000 on 32 percent of the plantations; and more than $5,000 on the remaining 30 percent. On 111 plantations the operator's gross cash income per unit was more than $8,000, and on 1 plantation in the Arkansas River area it amounted to more than $61,000. More than half of the plantations on which the operator's gross cash income was greater than $8,000 per unit were in the Upper Delta, the Red River, and the Arkansas River areas. About half the units on which the operator's gross cash income was less than $500 were in the Lower Delta area and in the renter majority counties of the Black Belt area.

[6]The reader is warned against taking this figure as the operator's share of the plantation gross cash income, which can be derived from Table 24. As has been explained the figures are not exactly comparable because of intra-plantation transactions.

Net Income

The gross cash income of the operator is, to a large extent, merely a reflection of the size of the plantation and the general type of working agreement which is in force between the operator and the laborers, croppers, or tenants on the plantation. On the other hand, the operator's net income, which is arrived at by deducting his expenses from his gross income, is more nearly a reflection of his managerial efforts. The relation between operator net income and net plantation income by areas is indicated in Figure 18.

Table 28—OPERATOR'S NET INCOME, BY AREAS, 1934
(Cotton Plantation Enumeration)

Area	Total Plantations Reporting[a]	Total		Cash		Home Use		
		Amount	Per Plantation	Amount	Per Plantation	Amount	Per Plantation	Percent of Total
Total	645	$1,659,082	$2,572	$1,491,842	$2,313	$167,240	$259	10.1
Atlantic Coast Plain	56	148,600	2,654	129,000	2,304	19,600	350	13.2
Upper Piedmont	40	68,399	1,710	56,959	1,424	11,440	286	16.7
Black Belt (A)[b]	112	163,794	1,462	132,210	1,180	31,584	282	19.3
Black Belt (B)[c]	99	134,765	1,361	108,035	1,091	26,730	270	19.8
Upper Delta	133	514,600	3,869	481,350	3,619	33,250	250	6.5
Lower Delta	49	84,692	1,728	77,342	1,578	7,350	150	8.7
Muscle Shoals	22	29,488	1,340	24,318	1,105	5,170	235	17.5
Interior Plain	30	70,952	2,365	60,362	2,012	10,590	353	14.9
Mississippi Bluffs	47	110,923	2,360	102,510	2,181	8,413	179	7.6
Red River	28	125,542	4,484	118,374	4,228	7,168	256	5.7
Arkansas River	29	207,327	7,149	201,382	6,944	5,945	205	2.9

[a] Data not available for one plantation.
[b] Cropper and other share tenant majority.
[c] Renter majority.

The average net income of the operators of the 645 plantations covered by this study was $2,572, of which approximately 10 percent, or $259 was in the form of products used for home consumption (Table 28). The operator's average net income ranged from $1,340 in the Muscle Shoals area to $7,149 in the Arkansas River area. The average net income of the operator in 7 of the 11 areas was less than the average for all areas. The proportion of the operator's net income made up of products used for home consumption varied from 2.9 percent of the total in the Arkansas River area to 19.8 percent of the total in the renter majority counties of the Black Belt area. In the Upper Delta, Red River, and Arkansas River areas, where the operator's income was highest, the value of home use production was especially small relative to net cash income. In all but two of the areas the value of the products used by the operators for home consumption averaged less than $300 per year.

The average operator's net income of $2,572 represents a return for his labor and capital as well as any remuneration which might be viewed as a reward for risk-bearing. It is, in brief, the end result of his year of farming operations. To apportion a part of it to his labor and another part to his capital investment is an arbitrary procedure. However, it is one commonly used and may be an aid in interpreting the net income figures shown in Table 28.[7]

The average net income of the operators of the 632 plantations for which complete data on investment were obtained was $2,576, and the average investment in these plantations was $28,694 (Appendix Table 11). Hence, the entire net income of the average operator represented a return of only about 9 percent on his investment, if no allowance is made for his labor and risk-taking or for depreciation or interest on production capital not borrowed (Appendix Table 36). If the average operator is allowed $500 per year for his labor, the remaining net income is equivalent to 7.2 percent on his investment. If $1,000 per year is allowed for the return to the operator's labor, the remaining net income is equal to a return of 5.5 percent on his investment. On either basis the highest return on the operator's capital was in the Atlantic Coast Plain area, where the operator's net income was a little higher than the average for all areas and where his investment was far below the general average (Appendix Table 36).[8] For the relation between operator net income and labor income see Figure 19.

A more commonly followed procedure in an analysis of this nature is to allow a given return on the investment and then attribute the remaining net income to the operator's labor. The interest on mortgages was approximately 6 percent per annum, and if this rate of return is allowed on the balance of the operator's investment, the average annual labor income is reduced to $855. Since the total net income of the operator in the Interior Plain area was only 5.1 percent on the capital invested, the procedure of allowing a 6 percent return on investment gives a minus labor income for the average operator in this area of $426 per year. This merely indicates that the average operator in this area had a total net income which was $426 less than an amount equivalent to 6 percent per annum on his investment. This was the only area in which the average operator's net income was less than a sum equivalent to 6 percent of his capital investment. Nevertheless, there were some plantations in every area on which the operator's net income did not yield 6 percent. For instance, 32 percent of the plantations

---

[7]The capital owned by the operator was not segregated from that rented on 13 of the 645 plantations; hence, for the purposes at hand the discussion must necessarily pertain to only 632 plantations. The data would not have been greatly changed, however, by including the figures for the other 13 plantations.

[8]This is explained by the fact that although all plantations included in the study derived 40 percent or more of their income from cotton production, a number of the sample plantations in the Atlantic Coast Plain area raised tobacco with a high income productivity per acre.

in the cropper majority counties of the Black Belt area, 39 percent in the Lower Delta, and approximately 40 percent in the Red River area were in this category.

A total of 181 plantations, or about 29 percent of the 632 for which complete records were obtained, failed to yield the operator a net income equivalent to a 6 percent return on his investment (Table 29). An additional 117 units failed to earn the operator a net income large enough to allow 6 percent on his investment and an additional return to his labor of as much as $500 per year. On the other hand, there were 114 units, or 18 percent of the total, on which the operator's net income was

Table 29—OPERATOR LABOR INCOME,[a] 1934
(Cotton Plantation Enumeration)

Labor Income	Plantations Reporting[b]		Total Labor Income	Average Labor Income
	Number	Percent		
Total	632	100.0	$540,145	$ 855
Loss $500 and over	98	15.5	-250,482	-2,556
Loss less than $500	83	13.1	-18,342	-221
Gain less than $500	117	18.5	26,882	230
Gain $ 500 to $1,000	98	15.5	70,624	721
Gain $1,000 to $1,500	67	10.6	83,375	1,244
Gain $1,500 to $2,000	55	8.7	94,694	1,722
Gain $2,000 to $2,500	26	4.2	59,017	2,270
Gain $2,500 to $3,000	23	3.6	63,649	2,767
Gain $3,000 to $3,500	19	3.0	61,116	3,217
Gain $3,500 to $4,000	9	1.4	33,590	3,732
Gain $4,000 to $4,500	4	0.6	17,258	4,315
Gain $4,500 to $5,000	3	0.5	13,869	4,623
Gain $5,000 and over	30	4.8	284,895	9,497

[a] Net income (cash and home use) less 6 percent on capital investment in land, buildings (excluding operator's residence, gins, and commissaries), animals, and machinery. Enumerators were instructed to "enter values at conservative market value, not low assessed value or high speculative value."

[b] Data not available for 14 plantations.

equivalent to a 6 percent return on his investment plus an additional amount of his labor and risk-bearing of $2,000 or more. Many of the plantations in the most successful groups were in the Arkansas River and Upper Delta areas, with a few scattered units in all except the Muscle Shoals area. The average labor income, after an allowance of 6 percent on capital, was $4,679 per unit for this group of the 114 most successful plantation operators, or more than 5 times as large as the average for all areas (Table 29).

There seems to be no definite relationship between operator labor income per crop acre and per acre value of land (Appendix Table 37). Of the plantations with an average per acre value of land of less than $10, 55 percent had an operator labor income per crop acre which represented a loss, or a gain of less than $2.50. Of those plantations with an average per acre value of land of $40 and over, 56 percent reported a loss, or a gain of less than $2.50 as the operator labor income per crop acre. This is probably accounted for by the fact that other factors than fertility enter into land values and by equalization of fertility by use of commercial fertilizer.

Net Cash Income

The operator's net cash income is perhaps more significant for comparative study than his total net income. The total net cash income of the operators of plantations averaged $2,313 per unit (Table 30), ranging from an average of $1,091 per planta-

Table 30—OPERATOR'S NET CASH INCOME, BY AREAS, 1934
(Cotton Plantation Enumeration)

Area	Total Plantations Reporting[a]	Average Net Cash Income
Total	645	$2,313
Atlantic Coast Plain	56	2,304
Upper Piedmont	40	1,424
Black Belt (A)[b]	112	1,180
Black Belt (B)[c]	99	1,091
Upper Delta	133	3,619
Lower Delta	49	1,578
Muscle Shoals	22	1,105
Interior Plain	30	2,012
Mississippi Bluffs	47	2,181
Red River	28	4,228
Arkansas River	29	6,944

[a] Data not available for one plantation.
[b] Cropper and other share tenant majority.
[c] Renter majority.

tion in the renter-majority counties of the Black Belt to a high of $6,944 per unit in the Arkansas River area. The net cash income of the average operator was lower in 8 of the 11 areas than the average for all areas. In five of the areas it was less than $2,000. On 43 of the plantations, the operator suffered a net cash loss during 1934, and on 151 plantations, or about

Table 31—OPERATOR'S NET CASH GAIN OR LOSS, 1934
(Cotton Plantation Enumeration)

Net Cash Gain or Loss	Plantations Reporting[a]		Net Cash Gain or Loss	
	Number	Percent	Total	Average
Total	645	100.0	$1,491,842	$2,313
Loss $500 and over	18	2.8	-26,081	-1,449
Loss less than $500	25	3.9	-4,086	-163
Gain less than $500	108	16.7	31,304	290
Gain $500 to $1,000	135	20.9	96,331	714
Gain $1,000 to $1,500	77	11.9	97,337	1,264
Gain $1,500 to $2,000	61	9.5	105,287	1,726
Gain $2,000 and over	221	34.3	1,191,750	5,393

[a] Data not available for one plantation.

one-fourth of the 645 reporting, the operator either had a cash loss or a net cash income of less than $500 per unit (Table 31). The areas in which the greatest proportion of plantation operators reported cash gains of less than $500 were the Lower Delta and the two Black Belts. However, the proportion of plantations on which the operator suffered an actual cash loss was not as great in those areas as in others. For instance, 13.5 percent of the plantations in the Arkansas River area and 12.5 percent of those in the Atlantic Coast Plain area failed to make a cash return for the operator as large as his current cash outlays for the 1934 crop; whereas, in the Lower Delta and the Black Belt areas less than 10 percent of the operators incurred a cash loss. Notwithstanding the heavy proportion of units in the Arkansas River area in which the operator sustained a loss, 21 of the 29 operators surveyed in this area had a net cash income of more than $2,000 per plantation. The Red River area and the Upper Delta also had large proportions of plantations on which the operator's cash income was more than $2,000 per plantation.

Table 32—SIZE AND CROPPING CHARACTERISTICS OF PLANTATIONS, BY OPERATOR'S NET CASH INCOME PER ACRE, 1934
(Cotton Plantation Enumeration)

Operator's Net Cash Income per Acre	Total Plantations Reporting[a]	Operator's Net Cash Income per Plantation	Acres per Plantation			Cotton Yield per Acre (pounds)	Percentage of Land in Crops	Percentage of Crop Land in Cotton
			Total	Crop	Cotton			
Total	645	$2,313	905	383	151	253	42.1	39.4
Loss $1.25 and over	6	-1,975	555	344	88	188	62.0	25.6
Loss less than $1.25	36	-509	1,720	410	143	211	23.8	34.9
Gain less than $1.25	166	631	1,070	281	76	203	26.3	27.0
Gain $1.25 to $2.50	143	1,575	879	375	135	233	42.7	36.0
Gain $2.50 to $3.75	93	2,493	823	431	178	271	52.4	41.3
Gain $3.75 to $5.00	62	3,666	836	505	249	266	60.4	49.3
Gain $5.00 to $6.25	37	3,851	685	435	179	321	63.5	41.1
Gain $6.25 to $7.50	32	3,912	559	388	175	267	69.4	45.1
Gain $7.50 to $8.75	29	6,598	821	513	284	346	62.5	55.4
Gain $8.75 to $10.00	17	4,699	540	385	172	293	71.3	44.7
Gain $10.00 and over	24	8,066	569	379	178	358	66.6	47.0

[a]Data not available for one plantation.

A special area-by-area study was made of the 43 plantations on which the operators suffered a cash loss, in order to find some common characteristic which might explain why they did not yield a return to the operator equivalent at least to his current cash outlay. The only common characteristic found, even within a given area, was that a relatively large proportion of the operators were non-residents and were engaged in some type of occupation other than farming. Although this relationship was not clear-cut, it did appear to be of some significance. On the other hand, factors such as size of plantation and high degree of specialization in cotton production, which were related to total plantation income, were not uniformly common characteristics of the plantations on which the operator incurred a cash loss.

When the operator's net cash income was computed on a per acre basis as shown in Table 32, it appeared that the average yield of cotton per acre and the percentage of crop land in cotton were the two most important factors related to net cash income. Most of the plantations on which the operator had a relatively high net cash income per acre were smaller than the average, but this was also true of the six plantations on which the operator suffered the greatest loss. Moreover, there appears to be practically no relationship between the crop acres per plantation and the operator's net cash income per acre.

For instance, on those units which yielded the operator a net cash income of $10 or more per acre, there was an average of 379 acres in crops, which is about the same as the average for all plantations studied. Most of the plantations which had a relatively high acreage in cotton made a higher than average net cash income per acre for the operator. This relationship, however, may not be significant and obviously is not uniform. The largest acreage of cotton per plantation was on those units which yielded the operator a net cash income of from $7.50 to $8.75 per acre, but the average acreage in cotton per unit for the plantations yielding the highest net cash income per acre was not significantly greater than the average for all plantations. On the other hand, every group of plantations on which the operator made a net cash income of more than $2.50 per acre had a higher than average yield of cotton and a higher than average proportion of crop land in cotton. Moreover, every plantation on which the operator suffered a loss, or had a relatively low net cash income per acre, had a relatively low per acre yield of cotton and, similarly, a low percentage of crop land in cotton.

TENANTS' AND LABORERS' INCOME

Most of the croppers and other share tenants on the plantations of the South do not buy their supplies and equipment or sell their products in the manner ordinarily followed by small farmers in other areas. They are usually advanced their seed, feed, fertilizer, and items for family living by the landlord, and they commonly turn their crops over to him for sale. Consequently, they are not familiar with the details of the nature or amount of their individual receipts and expenses. They usually know only the total amount of their seasonal indebtedness to the landlord and the amount of cash which they received or the amount which they still owed at settlement time. In addition, they are familiar with the few minor expenses and sales arising from transactions with persons other than the landlord.

Even though many of the tenants and croppers are unable to segregate strictly farm expenses from their household expenditures for subsistence, it is nevertheless possible to present data evaluating their net income.

The net income was calculated for tenants and croppers by the following method. To the cash after settling with the landlord were added: (1) advances for subsistence; (2) A.A.A. benefit payments; (3) the value of products used for home consumption;

(4) receipts from unshared sales of crops and livestock products; and (5) wages earned on the plantation.[9] From this sum was subtracted the unshared expenses borne by the tenant or cropper. Although this method of calculating net income departs somewhat from customary procedure, it nevertheless gives a more accurate and comparable picture for croppers and other share tenants than can probably be obtained by the more straightforward procedure of listing farming expenses and receipts and calculating net income by obtaining their difference. The latter procedure, however, is applicable and was used in the case of cash renters and laborers, who are not ordinarily advanced their subsistence by the operator and do not have a crop sharing rental agreement.

Net Income

The average annual net income of the wage hands, croppers, other share tenants, and renters in the 11 areas surveyed was $309 per family or $73 per capita (Appendix Table 38). However, there was great variation in the net income among the various groups in a given area, and also within a given tenure class among the various areas.

The average net income per family of the wage laborers was $180 for the year, and varied from $213 in the Arkansas River area to $70 in the Interior Plain. It is significant that the average net income per capita in the wage laborer group ranged in the various areas from $52 to $96 for the year. The average of $62 for all areas is about 17 cents per day. In every area, except the Lower Delta, the annual family income of the wage workers was less than that in any other status. On a per capita basis, however, the laborers' average net income was higher than the croppers' in four areas; higher than the share tenants' in two areas; and higher than the cash renters' in the same two areas (Lower Delta and Mississippi Bluffs areas) (Appendix Table 38).

More than half of the total number of families for which income data were obtained were share-croppers, whose average net income for all areas was $312 per family or $71 per capita. The croppers' average income per family and per capita was highest in the Atlantic Coast Plain area and lowest in the Lower Delta. In the latter area the croppers' average net income amounted to $38 per person, or slightly more than 10 cents per day. In the cropper majority counties of the Black Belt the average net income of the share-cropper families was slightly higher than that of the other share tenants. In all other areas the croppers

---

[9] The item "cash after settling" is the amount which the operator owed the tenant after dividing and selling the crops and deducting the tenant's share of farm expenses and the amount of his subsistence advance; or, obversely, it may be the amount which the tenant owes the operator at the end of the year.

had a lower net income, both per family and per person, than either the other share tenants or cash renters.

The share tenants had an average net income of $417 per family, or $92 per capita, the highest of any of the tenure groups. Five of the eleven areas reported more than this average, but only three—the Atlantic Coast Plain, Interior Plain, and Red River areas—reported as much as $110 a year or 30 cents per day as the net per capita income of share tenants.

A majority of the 650 cash renters were in the Black Belt and Lower Delta areas, which were among the poorest areas studied. The heavy concentration of cash renters in these areas resulted in the average net income of the renters for all areas ($354 per family) being considerably lower than the comparable average for other share tenants (Appendix Table 38). In most areas, however, the net income of the renters was higher than

Table 33—CROPPER AND OTHER SHARE TENANT NET INCOME[a] PER FAMILY, AND PERCENT OF CROP ACRES IN COTTON, 1934
(Cotton Plantation Enumeration)

Net Income Per Family	Plantations Reporting[b]		Percentage of Crop Acres in Cotton			
	Number	Percent	Less than 30	30 - 50	50 - 70	70 and over
Total	524	100.0	95	273	131	25
$ 50 to $100	7	1.3	1	4	2	
100 to 150	29	5.5	4	16	8	1
150 to 200	37	7.1	7	21	6	3
200 to 250	98	18.7	21	44	23	10
250 to 300	85	16.2	8	46	28	3
300 to 400	134	25.7	21	74	33	6
400 to 500	71	13.5	17	38	15	1
500 and over	63	12.0	16	30	16	1
Median income	$304		$331	$307	$297	$243

[a]Cash and home use.
[b]Tobacco plantations and cattle plantations excluded.

that of any other tenure group. The average net income of the renters was highest in the Upper Delta, where it amounted to $561 per family, or $146 per capita.

Variations in tenant net income are directly related to the quality of the land operated. Thus, the net income of croppers and other share tenants on land valued at $20 or more per acre was found to be definitely higher than the income of those on poorer land (Appendix Table 39).

Since it has been pointed out that plantation tenants often are not allowed by landlords to devote much land or time to food crops, it may also be significant to note how net income of croppers and other share tenants varies with the ratio of cotton acres to all crop acres on the plantation. It is evident from Table 33 that there is a tendency for croppers and other share tenants on plantations with a small percentage of crop land in cotton to have a relatively high net income. As the percentage of crop acres devoted to cotton increases, the net income of croppers and other share tenants decreases, the median net income per family on plantations having less than 30 percent of their crop acreage planted in cotton being $331, as compared with $304 for all plantations and $243 for plantations devoting 70 percent or more of their crop acreage to cotton. The fact that income decreases as cotton acreage increases is because the higher incomes of plantation families are those which include large amounts of home use production (Appendix Table 40).

Net Cash Income

Since the net income of croppers and other share tenants includes the value of advances for subsistence made to them by the

Table 34—CROPPER AND OTHER SHARE TENANT NET CASH INCOME, BY AREAS, 1934
(Cotton Plantation Enumeration)

Area	Croppers				Other Share Tenants			
	Total Families Reporting[a]	Net Cash Income			Total Families Reporting[b]	Net Cash Income		
		Per Family	Per Person	Percent of Total Net Income		Per Family	Per Person	Percent of Total Net Income
Total	2,873	$122	$28	40.1	705	$202	$44	47.4
Atlantic Coast Plain	212	255	43	49.4	16	426	70	51.1
Upper Piedmont	124	104	20	31.1	52	170	31	39.9
Black Belt (A)[c]	404	127	25	38.4	62	106	22	33.9
Black Belt (B)[d]	232	71	13	29.4	23	119	18	29.1
Upper Delta	923	138	35	43.0	272	213	51	50.9
Lower Delta	136	42	10	27.7	49	51	12	23.5
Muscle Shoals	29	137	30	40.6	46	211	45	42.8
Interior Plain	172	109	24	33.5	48	215	45	39.4
Mississippi Bluffs	257	75	19	33.5	61	149	31	53.6
Red River	125	129	35	42.2	29	420	88	60.1
Arkansas River	259	109	33	45.2	47	206	60	72.7

[a] Data not available for 13 families.
[b] Data not available for 11 families.
[c] Cropper and other share tenant majority.
[d] Renter majority.

landlord during the crop-making season and of products used for home consumption, it is important to show the proportion of their total net income which they receive in the form of cash. It has often been claimed that most of the share-croppers and tenants of the South receive practically no cash income, and are very largely dependent upon the credit advances, usually called "furnish", made by the landlord. Although approximately 10 percent of the cropper families and about 17 percent of the share tenant families were in debt to the plantation operator at the end of the season (Table 22),[10] and, hence, received very little cash from their farming operations during the year, it was found that the average cropper family had a cash income

[10] See chapter V.

of $122, or $28 per person and the average share tenant family had a cash income of $202, which was equivalent to $44 per person. Approximately 40 percent of the croppers' net income, and about 47 percent of the net income of the share tenants, was in the form of cash (Table 34).

Among the cropper group the highest net cash income, both per family and per person, was in the Atlantic Coast Plain area, where the net cash income of $255 per family was over two-fifths of their total net income. In the Lower Delta, the average net cash income per family and per person among the croppers reached the extreme low of $42 and $10, respectively, this being only 28 percent of the total net income. The average net cash income per family and per capita was higher for the other share tenants than for the share-croppers in all areas except the cropper majority Black Belt. As was true of the croppers, the share tenants in the Lower Delta ranked at the bottom of the list both in the amount of net cash income received and in the proportion it was of their total net income. The average net cash income of the share tenants in this area was only $51 per family.

About three-fourths of the net cash income of both the tenants and the croppers was accounted for by "cash after settling." As has been previously explained, "cash after settling" refers to the difference between the value of the cropper's or tenant's part of the crop and the amount which he owed the operator for farming supplies, subsistence advances, and interest. The average cropper had a cash income of $21 in extra wages[11] and $8 in A.A.A. benefit payments. The average share tenant received $17 in cash from each of these sources. In both tenure groups the net receipts from sales of unshared crops and livestock products were almost negligible.

Only in the Atlantic Coast Plain and Muscle Shoals areas did the croppers receive an average of more than $100 per family after their share of farming expenses and their subsistence advances had been deducted from the value of their part of the crops produced. In the former area, the 212 croppers from whom complete records were obtained received an average of $218 per family in the form of cash after settling; however, 10 percent of the group either "broke even" or were in debt from their farming operations. In the Lower Delta area the average amount of cash after settling received by the cropper families was $33, and only 70 percent of the families had any cash due them. Approximately 17 percent of the families in this area "broke even", and 13 percent suffered a loss. The cash after settling received by share tenant families was considerably larger in amount than among cropper families, but a larger proportion merely "broke

---

[11] If the cropper performs plantation duties other than those incident to producing the shared crop, he receives extra pay.

even" or lost. In the Lower Delta only 53 percent of the share tenants had any cash due them at settlement time, and the average amount for all the tenant families in the area was only $28 per family. The highest amount of cash after settling per family among the share tenants was in the Atlantic Coast Plain area, where the 16 families averaged $373 although 25 percent of them received no cash at settlement time.

The following summary shows the source of the average net income of wage hands, croppers, share tenants, and cash renters covered by the survey. As is evident from these figures, over

Table 34-A—NET INCOME OF PLANTATION FAMILIES, BY TYPE, 1934
(Cotton Plantation Enumeration)

	Wage Hands	Croppers	Other Share Tenants	Cash Renters
Total net income	$180	$312	$417	$354
Cash items	148	122	202	196
Cash after settling	–	91	152	–
Wages	148	21	17	–
A.A.A. payments	–	8	17	26
Unshared sales	–	2	16	170
Non-cash items	32	190	215	158
Subsistence advances	–	85	70	–
Home use production	32	105	145	158

80 percent of the income of the wage hands was in the form of cash wage payments. On the other hand, the share-croppers received only about 40 percent of their income in the form of cash, and approximately 60 percent was made up of subsistence advances by the landlord and commodities produced for home consumption. Almost half of the share tenants' income was in the form of cash, most of which was "cash after settling." The subsistence advance made by the landlord to share-croppers and other share tenants was considerably less than the value of the products which they produced for their home consumption, and amounted to approximately one-fourth and one-sixth, respectively, of their total net incomes. About 55 percent of the net income of the cash renters was in the form of cash, and was primarily from the sale of crops and livestock products. The balance of their income was in the form of A.A.A. payments and products for home use (Table 34-A).

MUTUAL DEPENDENCE OF TENANT AND LANDLORD

It has been brought out in the preceding sections that 70 percent of the plantation income was from sale of crops and that only 30 percent included unshared items, such as miscellaneous livestock and livestock product sales, and products consumed by the plantation families. Thus, 70 percent of the plantation income is shared income (except where it is produced by the landlord with hired labor). It is almost exclusively upon crops that the landlord and tenant are mutually dependent for cash.

Obviously then the landlord and tenant are mutually dependent upon the factors which affect crop production and price. Less capable of objective measurement is the managerial ability of the landlord and the energy and ability of the tenant. Both of these, however, are factors affecting production and in these respects the joint cash income of landlord and tenant are interdependent.

Appendix Table 41 indicates the extent to which tenant income and landlord income vary together. There is a fairly steady rise in tenant income as operator income per crop acre rises.

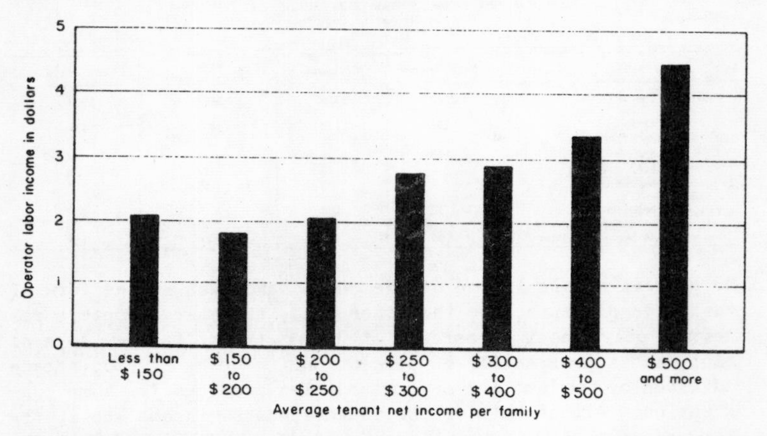

FIG. 20 - MEDIAN OPERATOR LABOR INCOME PER CROP ACRE BY AVERAGE TENANT NET INCOME PER FAMILY 1934

Source: Appendix table 41

AF - 1427, W.P.A

On the plantations yielding the landlord $15 per acre and better, nearly 90 percent of the tenants were earning $300 or more per family. On the plantations where landlords showed an actual loss one tenant in every six netted less than $200 and only about 46 percent made $300 or more. Figure 20 shows the steady rise in operator labor income per acre as tenant net income increases. On plantations where tenants received an average of $500 or more net income the median operator labor income per acre is about two and one-half times that for plantations on which tenants averaged only $150 to $199.

The efficiency of plantation operations seems to vary with the type of rental agreement entered into between the landlord and the plantation families, which determines to a large extent the amount of supervision provided by the landlord. This relationship becomes apparent in an analysis of cotton yield per acre by tenure status of the family (Table 35). Wage hands and croppers had an average cotton yield of 261 and 260 pounds per acre respectively, while the figure for share tenants was only

Table 35—COTTON YIELD PER ACRE, BY TENURE STATUS OF FAMILIES, BY AREAS, 1934
(Cotton Plantation Enumeration)

Area	Total Plantations Reporting[a]	Yield per Acre (pounds)			
		Total	Wage Hands	Croppers	Other Share Tenants
Total	602	257	261	260	244
Atlantic Coast Plain	55	293	293	284	258
Upper Piedmont	39	245	309	239	194
Black Belt (A)[b]	105	249	271	257	164
Black Belt (B)[c]	81	243	261	227	267
Upper Delta	132	297	264	301	305
Lower Delta	37	223	264	212	209
Muscle Shoals	21	246	288	248	233
Interior Plain	30	194	239	192	191
Mississippi Bluffs	45	260	238	278	206
Red River	28	234	213	247	240
Arkansas River	29	220	268	212	184

[a]Data not available for 44 plantations.
[b]Cropper and other share tenant majority.
[c]Renter majority.

244 pounds. In 7 of the 11 areas this order was followed. The explanation for the better results obtained by wage hands and croppers may lie partly in the closer supervision given to their operations, and partly in the fact that a landlord may have a tendency to assign the best land to such families, since he

Table 36—AVERAGE INCOME OF PLANTATIONS, BY TYPE OF OPERATION, 1934
(Cotton Plantation Enumeration)

Item	Plantations Operated Exclusively by					
	Wage Labor		Croppers and Other Share Tenants		Renters	
	Total	Average	Total	Average	Total	Average
Number of plantations	38		546		47	
Plantation total gross income[a]	$312,000	$8,211	$5,712,000	$10,462	$45,800	$ 974
Plantation total expenses[b]	186,300	4,903	2,089,000	3,826	17,600	374
Plantation net income	125,700	3,308	3,623,000	6,636	-	-
Tenant net income[c]	12,000	316	2,141,000	3,921	-	-
Landlord net income	113,700	2,992	1,482,000	2,714	28,200	600
Six percent return on investment	56,700	1,492	981,000	1,797	48,200	1,026
Landlord labor income	57,000	1,500	501,000	918	20,000	-426

[a]Includes cash income and home use production of landlord, croppers, share tenants, and wage hands; excludes all income of renters; includes landlord income from cash rent.
[b]Includes expenses of landlord and tenants (shared and unshared) but excludes expenses of renters.
[c]Includes cash and home use production; excludes cash income of wage hands.

receives a larger share of their crop than he receives from other share tenants.

It does appear, however, that operators using wage hands exclusively had a larger labor income than those who employed mainly croppers and share tenants, the averages being $1,500 and $918, respectively (Table 36). Those operating their plantations exclusively with renters averaged a labor income of minus $426, this type of operation obviously being much less profitable.

Chapter VII

TENANT'S STANDARD OF LIVING

Although the present study did not concern itself directly with the details of the tenant's standard of living, much of the material in other chapters bears on tenant incomes and expenditures. It is the purpose of this chapter to draw these facts together with the findings of other studies which reflect on the tenant's way of life. These facts are concerned principally with the free services of the plantation, such as fuel and housing; low cash incomes; the custom of paying part of the income in advance in supplies instead of money; and the small amount of foodstuff produced for home use.

Fundamental to the understanding of the living habits of tenants is the realization that they exercise a relatively limited choice in determining these habits, and have been supervised for so long that if they did have a freer choice they would not have the knowledge of other ways of living essential to change. The system of agriculture determines, first of all, that they devote themselves almost exclusively to cotton, supplemented by enough corn to feed the plantation animals. The landlord determines what sort of house the tenant shall live in and what the amount and characteristics of the monthly "furnish" of foodstuffs shall be. If the tenant has lived under these conditions for a number of years, not only cultivating his land under close supervision but also arranging his household budget under equally rigid oversight, the loss of initiative and self-reliance is marked.

A peculiarity of tenant income is the irregularity of its distribution over the year. The subsistence advance of the tenants in this study averaged about $13 a month for 7 months (Table 21). This represents approximately a third of the tenant's income[1] and usually begins in January or February. During the spring and summer months the advance constitutes practically the only family outlay. The other two-thirds of the tenant's income is derived from home use products and the cash balance for his share of the crops sold. These latter two items in 1934 amounted on the average to about $200 for the croppers and $300 for the other share tenants in this study. Sale of crops usually

---

[1]See chapter VI.

takes place in the fall and early winter months. Much of the cash is immediately used for payment of extra plantation debts such as doctors' bills and for annual purchases such as clothing and household equipment. The home use products, predominantly pork, corn meal, syrup, sweet potatoes, and cow peas, are also available primarily in the fall, and since storage facilities are limited these products must be consumed within a short time. Thus, one-third of the income is spread over two-thirds of the year and two-thirds of the income is expended in one-third of the year.

HOUSING AND FUEL

No attempt was made in income calculations of this study to estimate the value of plantation perquisites, such as house, fuel, game, berries, and nuts, which are free to the tenant. The use of the land provided in the landlord-tenant agreement carries with it these privileges, none of which except house and fuel is of appreciable value. It is difficult to place a money value on the fuel used by tenants because it consists entirely of wood and the price of wood in town is largely conditioned by the cost of cutting and hauling, services which are performed by the tenant for himself, so that he receives only the value of the wood.

The housing furnished tenants in the Cotton Belt can, however, be evaluated by objective measures, notwithstanding the fact that these measures must be interpreted with the realization that actual physical as well as cultural requirements differ for various areas and that few generally accepted minimum or optimum standards exist. From any measure adopted, however, it appears that the housing for cotton tenants is below the level of any other large segment of the Nation's population.

Value of Housing

According to 1930 Census reports, the average appraisal value of farm dwellings in the seven southeastern cotton States—Alabama, Arkansas, Georgia, Louisiana, Mississippi, North Carolina, and South Carolina—is lower than that for the United States as a whole or for any geographic division thereof (Appendix Table 42 and Figure 21). In considering the Census figures the reader is reminded that they represent the farmers' estimates of the value of their buildings and "are probably somewhat less satisfactory than the figures for the total real estate value."[2]

---

[2] *Fifteenth Census of the United States:* 1930, Agriculture Vol. I, p. 2.

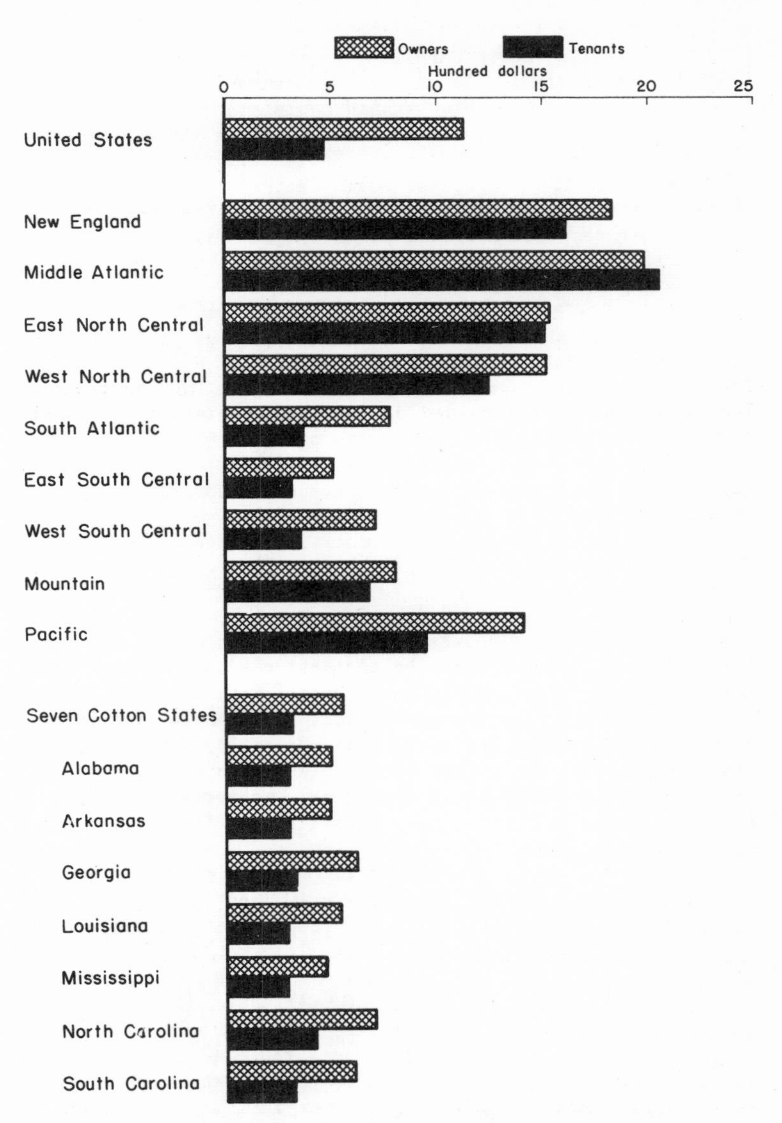

FIG. 21 - MEDIAN VALUE OF FARM DWELLINGS, BY TENURE, 1930

Source: Fifteenth Census of the United States: 1930, special release

AF-1495, W. P. A.

The seven southeastern cotton States fall into three geographic divisions as listed in the United States Census:

South Atlantic—North Carolina, South Carolina, and Georgia.

East South Central—Alabama and Mississippi.

West South Central—Arkansas and Louisiana.

The average values of farmers' dwellings in these areas are the lowest in the United States: South Atlantic, $783; West South Central, $584; and East South Central, $503. Moreover, in every instance, the States under survey have the lowest values within their respective areas. The average for all rural dwellings in these seven States is $467, including the homes of both owners and tenants. This may be contrasted with average values of dwellings in the other regions: Middle Atlantic, $2,237; New England, $2,218; East North Central, $1,657; Pacific Coast, $1,617; West North Central, $1,559; and Mountain, $989. Although the average value of dwellings in the Mountain States is lower than in any region outside of the South, it nevertheless is more than twice the average value of dwellings in the seven southeastern cotton States.

Table 37—AVERAGE VALUE OF FARM DWELLINGS IN SEVEN SOUTHEASTERN COTTON STATES, BY COLOR AND TENURE OF OPERATOR, 1930

State	Average Value					
	White			Negro		
	Total	Owners	Tenants	Total	Owners	Tenants
Seven cotton States	$594	$ 815	$380	$265	$373	$245
Alabama	512	721	337	220	348	194
Arkansas	464	615	325	233	355	213
Georgia	582	851	396	265	366	250
Louisiana	603	862	354	242	371	221
Mississippi	523	763	291	264	350	252
North Carolina	749	906	513	383	456	358
South Carolina	757	1,083	454	262	348	238

Source: *Fifteenth Census of the United States: 1930*, Agriculture Vol. IV, Table 30, and Vol. II, County Table 1, Supplemental for the Southern States.

Median values for dwellings of farm owners in the South Atlantic, West South Central, and East South Central regions were $782, $711, and $512, respectively. For tenant dwellings, the three southern regions had median values of $374, $361, and $314, respectively. For tenants in other areas, the Mountain States ranked lowest ($682) and the Middle Atlantic States highest ($2,058). The Middle Atlantic States also had the highest median value for owners ($1,986) (Appendix Table 42).

In the seven cotton States under consideration, the median value of tenants' houses ranged from $291 in Mississippi to $417 in North Carolina. This means that if the value of tenant dwellings was capitalized at 6 percent, the tenant received as a perquisite free rent to the amount of from $18 to $25 per year.

The 1930 Census reports indicate that for the seven States under consideration, owners' dwellings were evaluated, on the average, almost twice as high as those of tenants. The difference between the value of owners' and tenants' dwellings was much less in the case of the Negro than of the white farm families.

ONE OF THE WORST OF THE TENANT HOUSES

For the seven States as a whole the dwellings of the white families were appraised at twice the value of those of the Negroes (Table 37 and Figure 22).

Types of Dwellings

As shown by a Nation-wide farm housing study made by the Bureau of Home Economics in cooperation with the Civil Works Administration in 1934,[3] the majority of the houses in the seven

---

[3]In 1934 the Bureau of Home Economics, United States Department of Agriculture, and the Civil Works Administration cooperated in conducting a Farm Housing Survey. In the seven States, Alabama, Arkansas, Georgia, Louisiana, Mississippi, North Carolina, and South Carolina, 10 percent of all farms were included in the survey. An attempt was made to obtain a representative sample including roughly 10 percent of all white owners, white tenants, Negro owners, and Negro tenants.

FIG. 22

AVERAGE VALUE OF FARM DWELLINGS

IN SEVEN SOUTHEASTERN COTTON STATES AND THE UNITED STATES
1930

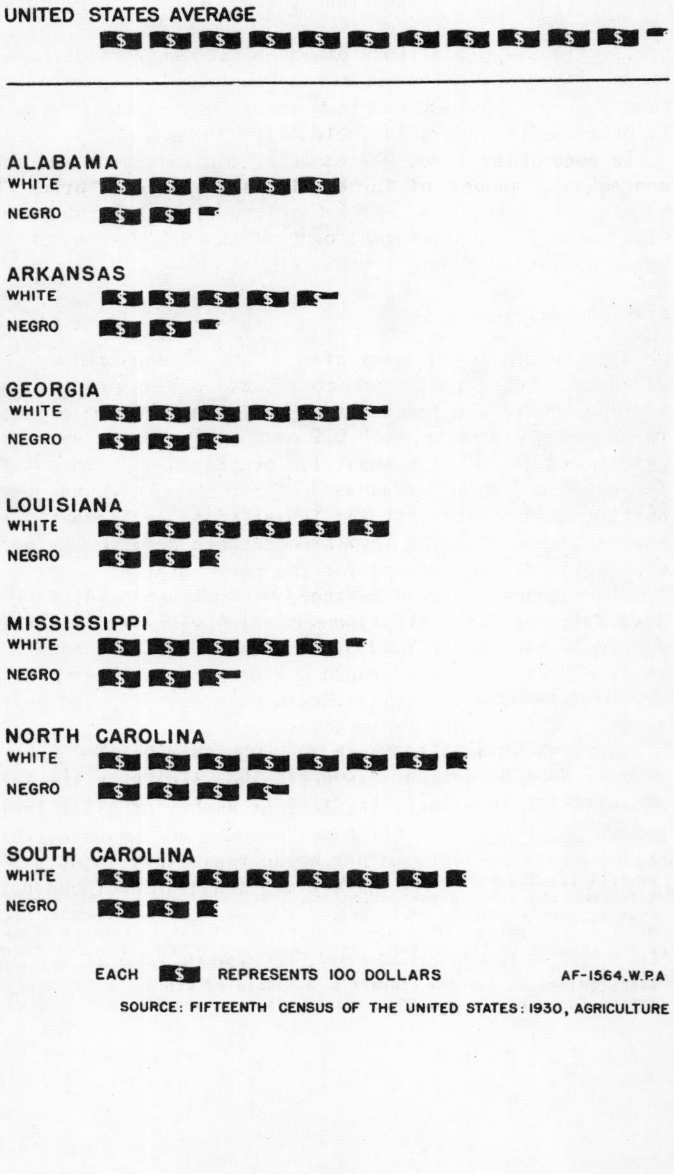

UNITED STATES AVERAGE

ALABAMA
WHITE
NEGRO

ARKANSAS
WHITE
NEGRO

GEORGIA
WHITE
NEGRO

LOUISIANA
WHITE
NEGRO

MISSISSIPPI
WHITE
NEGRO

NORTH CAROLINA
WHITE
NEGRO

SOUTH CAROLINA
WHITE
NEGRO

EACH $ REPRESENTS 100 DOLLARS

AF-1564.W.P.A

SOURCE: FIFTEENTH CENSUS OF THE UNITED STATES: 1930, AGRICULTURE

cotton States were unpainted frame dwellings.[4] Houses of this type were most prevalent in Louisiana and Mississippi where they constituted almost two-thirds of owners' houses and more than four-fifths of tenants' houses. Ninety-three percent of the Negro tenants in Louisiana lived in this type of dwelling.

Mississippi had the lowest proportion of painted frame houses, followed by Louisiana, while North Carolina had the highest proportion (Appendix Table 44). Except for Arizona, where many of the houses were adobes, fewer houses were painted in these seven southeastern States than in any other State included in the survey.

The highest proportions of log houses were in Alabama among Negro owners and white tenants (approximately 7 percent). Earth huts were virtually absent in the South except among Negro owners in Mississippi (Appendix Table 44).

In none of the seven States under consideration was there an appreciable number of farmers living in stucco, brick, stone, or concrete dwellings. North Carolina with the highest proportion had only 1.9 percent of owners' and 0.6 percent of tenants' homes of these types.

Size of Dwellings

With regard to adequacy of housing as measured by the size of house and number of regular occupants per room, North Carolina and South Carolina ranked highest with an average of about 5.7 rooms per owner house, with 0.9 occupant per room, and 4.5 rooms per tenant house with about 1.2 occupants per room (Appendix Tables 45 and 46). Arkansas had the fewest rooms per owner dwelling and Mississippi the fewest rooms per tenant dwelling. However, the number of regular occupants per room in each case was not above the average for the seven States.

Negro tenants lived in the most crowded conditions in all States except Mississippi, where Negro owners and white tenants were somewhat more crowded. For the seven States as a whole, white owners' houses averaged 0.9 occupant per room; white tenants' houses, 1.2 occupants; Negro owners', 1.2; and Negro tenants', 1.4 (Appendix Table 46).

North Carolina and South Carolina had on the average the largest number of bedrooms per house, and, with the exception of tenants in Arkansas, the largest number of other rooms (Appendix Table 47). In all cases owners and tenants, white and Negro, had more bedrooms per house than other rooms.

Kirkpatrick[5] reports the average number of bedrooms per house

---

[4]For number of farm houses surveyed, see Appendix Table 43.

[5]Kirkpatrick, E. L., *The Farmer's Standard of Living*, U. S. Department of Agriculture, Bulletin 1466, 1926, p. 21.

for selected areas to be as follows: Southern States, 3.0; Northern States, 3.3; New England States, 4.5.

Equipment of Dwellings

In all States in the southeastern cotton area the majority of white owners reported screens, the highest proportion being in Arkansas with 80.7 percent, the lowest in Georgia and Louisiana, with 59.6 and 59.4 percent, respectively (Appendix Table 48). In every State, Negro tenants had the lowest percentage of all groups reporting screens. In every State except Arkansas.

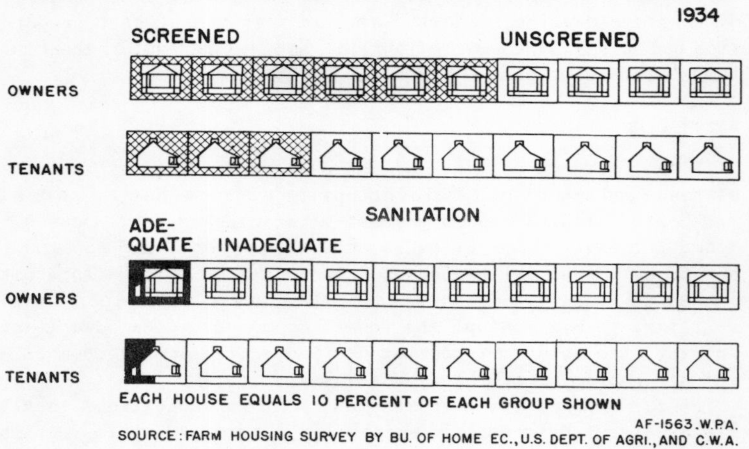

screens were reported by less than one-fourth of the Negro tenants. The proportion of white tenants with screens was considerably greater. Negro owners reported more screens than Negro tenants, but considerably fewer than white owners and tenants (Figure 23).

Wells furnished a source of water for over 80 percent of both owner and tenant dwellings in the cotton States (Appendix Table 49). About 60 percent reported wells, dug or bored, and 20 percent reported wells, drilled or driven. A slightly higher proportion of owners than of tenants reported wells. This was due primarily to the high percentage of white owners using wells in two States, Alabama and South Carolina.

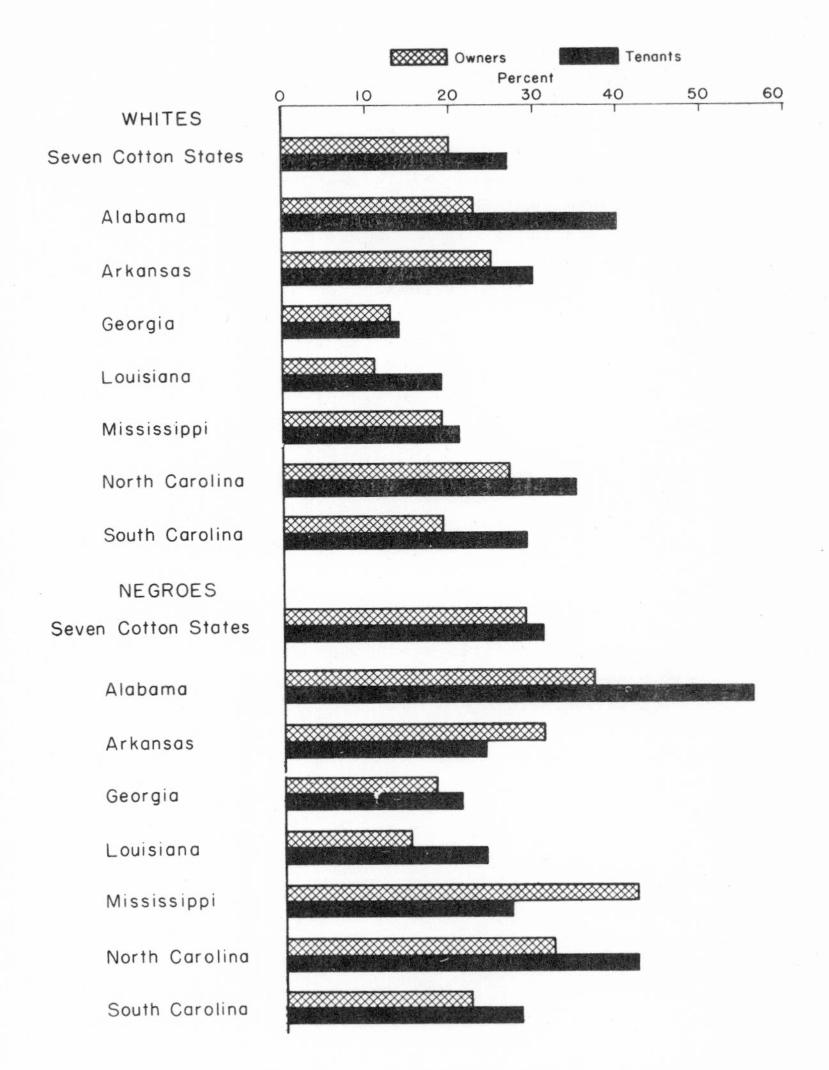

FIG. 24 – PERCENT OF FARM DWELLINGS IN SEVEN SOUTHEASTERN COTTON STATES WITHOUT SANITARY FACILITIES, 1934

Source: Farm Housing Survey by Bureau of Home Economics, U.S. Department of Agriculture in cooperation with Civil Works Administration

AF–1497, W P A

The use of spring water was reported by 11 percent of the owners and 10 percent of the tenants in the cotton States. Six percent of the owners' and four percent of the tenants' homes in the seven States were equipped with cisterns (Appendix Table 49). The use of stream water was reported for an average of less than 1 percent of the dwellings in the seven States, the highest proportion reported being among Negro tenants in Louisiana. Nevertheless, true percentages may be higher in some States than the survey data indicate.[6]

The most critical problem related to southern rural housing is the lack of sanitary facilities (Figures 23 and 24). The data now available indicate the seriousness of conditions, not only for Negroes and white tenants but for white owners as well. Compared to other classifications, white owners were somewhat better off, but 67 percent of them had only unimproved outhouses and 20 percent had no facilities whatsoever. Thus, 87 percent of the white owners had available only the poorest of conveniences, or none at all (Appendix Table 50).

While 68 percent of the white tenants had only unimproved outhouses, an additional 27 percent had no facilities. Among the Negro owners 67 percent had unimproved outhouses with 29 percent having no facilities, while 67 percent of the Negro tenants had unimproved outhouses and 31 percent had no facilities.

The proportion of dwellings equipped with indoor flush toilets was 4.6 percent for white owners, 1.0 percent for white tenants, 0.2 percent for Negro owners, and 0.1 percent for Negro tenants. Practically no dwellings were equipped with indoor chemical toilets, but a small percentage of the homes had improved outdoor toilets.

Over 90 percent of both owner and tenant families cooked on wood or coal stoves (Appendix Table 51). Less than 8 percent of owners' and 3 percent of tenants' homes were equipped with kerosene or gasoline stoves. Few of the white owners and tenants used gas or electric stoves while such stoves were not found among Negroes, either owners or tenants, throughout the cotton States.

For owners, a greater number of stoves were reported than there were houses surveyed. This duplication amounted to 3.5 percent of all owner dwellings, indicating more than one stove

---

[6] It should be noted that there is a certain amount of duplication and omission involved in the above figures, 2 percent more sources of water supply being reported for owners than there were owner houses surveyed in the area. In Louisiana there were 13 percent more sources than there were users who were reported. On the other hand, no State except Louisiana and North Carolina reported as many sources of water for tenants as there were tenant houses surveyed. In Arkansas, 9 percent of all tenants and 17 percent of Negro tenants gave no report on water supply. In Mississippi, 8 percent of all tenants and 11 percent of Negro tenants gave no report. It may be assumed that some of these families obtained water either from wells of neighbors or from streams.

per house in some instances. On the other hand, fewer stoves were reported for tenants than there were tenant houses surveyed. The scarcity of stoves was found only among Negro tenants, chiefly in Mississippi, where over one-fourth reported no cooking facilities. The proportions of Negro tenants without stoves in the other States were less significant.

CLOTHING

The clothing of tenant families is usually purchased annually except for odd items. This purchase is made at "settling" time if the cash is available to the tenants. If the tenant has made no profit on his operations or if he is unable to obtain further advance credit, his last year's wardrobe must suffice. Clothing purchased is of the coarsest, crudest character—denim overalls for the male members and cheap cotton goods for the female members. Brogan shoes, no socks, and homemade underwear, if any, are the rule. Sometimes the annual purchases for a large and prosperous family run as high as $100, or slightly more, but more often the amount is far less (Table 38). Often a lack of sufficient warn clothing prevents children from going to school and adults from attending church or other public gatherings.

FOOD

Dietary standards of tenants are as low as housing standards. It has been pointed out that the average tenant lives on an advance of about $13 per month for two-thirds of the year. This amount is adjusted by the landlord with some consideration for the size of the tenant's family, which averages about 4.4 persons, but more often the controlling factor is the amount which the landlord can borrow to finance these advances, and the previous record of the tenant in producing enough to repay these advances.

Not all of the $13 a month is used for food, since such miscellaneous purchases as kerosene, tobacco, medicine, and clothing, which cannot be postponed, must also be charged against this sum.

A detailed examination of the commissary accounts for 25 Arkansas tenants[7] in connection with the present survey gives a general picture of the purchasing practices of tenants under the subsistence advance. Table 38 indicates by items the outlay for food and the character of food items purchased. It will be noted that only staple articles are purchased, flour, lard,

---

[7]Examination of 25 accounts might be considered an inadequate basis for study were it not for the fact that tenants' consuming practices are so uniform throughout the South that the conclusions based on these 25 cases would not be greatly modified if a much larger sample were taken.

and meat accounting for most of the expenditures. The meat in this case is almost universally fat salt pork.

The most striking factor about these outlays is that no diminution of the amounts spent for food is reported for the summer months when it would be possible to supplement the family table with fresh vegetables. On the other hand, the reduction comes in the winter months when the crops have been sold and the tenant is living largely on cash purchases which are usually of the same type and grade as his credit purchases.

Table 38—ANNUAL COMMISSARY PURCHASES OF 25 TENANTS IN ARKANSAS, BY COMMODITIES, 1933

Commodity	Percent of Total Purchase
Total	100.0
Total food	64.4
Flour	23.3
Lard	12.1
Meat	9.1
Sugar	5.5
Condiments	5.4
Coffee	2.5
Molasses	1.7
Miscellaneous food	4.8
Clothing	14.2
Medicine	3.3
Tobacco	5.5
Miscellaneous household items	12.6

Source: Special tabulation for cotton plantation survey.

Gardens would, of course, be the best means of diversifying the diet of the family, but the practice of tending a garden is foreign to the habits of most tenants. Since the garden is not a shared operation, the only interest which the landlord has in the tenant's garden is the extent to which the production of foodstuffs will reduce the amount which he must lend the tenant for subsistence. The landlord is, therefore, not always willing to advance money for seed and fertilizer or to provide for the use of an animal for a tenant's garden. On the other hand, many landlords who attempt to encourage gardening among their tenants meet with opposition bred by a lifetime of cultivating only cotton and corn. Gardens found on plantations in this study were usually so poorly tended that the resulting small production in most cases could hardly be assigned an appreciable value in the tenant's budget. Under these circumstances, the canning of garden produce is very rare.

Aside from garden products, four field crops in the South can, and often do, contribute to the larder of the family. These are corn, ground for meal; cane, ground for syrup; sweet potatoes; and cow peas. Since these crops are not so marketable as cotton the landlord has no particular interest in financing fertilizer purchases for them and consequently the yields are usually low.

Domestic animals can also contribute to the diet of tenant families to a much larger degree than they now do. Table 39 indicates the extent to which the tenants interviewed owned domestic animals. Most of them had a few chickens, but there were some without even this small stock. Flocks were so limited on the average that they could contribute little in the way of either eggs or meat to the family food supply. Pigs were reported by 80 percent of the tenants and the pork produced was one of the largest items in the budget of home produced foods. However, pork is perishable unless cured and stored with more skill and equipment than is at the disposal of the average tenant.

Table 39—NUMBER OF COWS, PIGS, AND POULTRY REPORTED BY PLANTATION FAMILIES, BY TENURE, 1934
(Cotton Plantation Enumeration)

Tenure	Total Families	Families Reporting Cows		Cows		Families Reporting Pigs		Pigs		Families Reporting Poultry		Poultry	
		Number	Percent of Total Families	Number	Average per Family Reporting	Number	Percent of Total Families	Number	Average per Family Reporting	Number	Percent of Total Families	Number	Average per Family Reporting
Total	4,255	2,347	55.2	5,182	2	3,392	79.7	12,904	4	3,825	89.9	77,410	20
Croppers	2,886	1,310	45.4	2,135	2	2,219	76.9	7,633	3	2,531	87.7	46,358	18
Other share tenant	716	530	74.0	1,291	2	610	85.2	2,684	4	671	93.7	17,367	26
Renter	653	507	77.6	1,756	3	563	86.2	2,587	5	623	95.4	13,685	22

Consequently, the pork products are also usually consumed almost entirely in the late fall and early winter months.

One of the most serious diet deficiencies in the South is the absence of milk and butter from the tables of a large proportion of the tenant families. Only 55 percent of tenants had cows, a great proportion not having the knowledge or the energy to care for these animals or the money to purchase them. For those families who do have cows, milk is, of course, one of the chief items of home production.

Fewer croppers reported livestock and those with livestock had smaller numbers than was the case with other tenant groups. Only 45 percent of the croppers reported cows in comparison with 74 percent of the other share tenants and 78 percent of the renters. Not only did fewer croppers have pigs but they averaged only three pigs per family while other share tenants reported four and renters five (Table 39).

There is some indication from the 1935 Agricultural Census that one effect of the A.A.A. was to add somewhat to the production for home use in the South. It is impossible at this date to say whether this diversification has taken place primarily among the owners. There was substantial increase in the Southeast from 1930 to 1935 in the number of cattle and swine on farms.[8]

[8] *United States Census of Agriculture: 1935.*

This came about largely because the acreage retired from cotton production was available for pasture and home use production, undoubtedly increasing the land used for feed and forage crops and for grazing.

The measurement of the nutritional values of the meager tenant diet has not been carried out on an extended scale. No adequate study has been made covering a large segment of the South nor comparing the South to other areas. One study based on a small number of representative families in Mississippi[9] indicates that these families were provided with scarcely more than the actually necessary quantities of protein and phosphorus. Calcium was well provided (.69 grams used as standard); iron was not provided in sufficient quantities; vitamin A and B requirements were probably met, and vitamin C was perhaps low in the fall, winter, and early spring.

COMMUNICATION FACILITIES

Communication facilities form one of the best indices for comparing relative standards of living. In general, telephones, automobiles, and magazines are found to be least common in sections of the South with high percentages of farm tenancy. Table 40 indicates the lack of such modern methods of communication in the areas surveyed.

Table 40—TENANCY, TELEPHONES, PASSENGER CARS, AND MAGAZINES, BY AREAS,[a] 1930

Area	Percent Tenancy	Inhabitants per Telephone	Inhabitants per Passenger Car	Persons per National Magazine
Northern Cotton and Tobacco[b]	56	57	10	18
Southern Cotton and Tobacco[c]	60	61	14	23
Upper Piedmont[d]	63	41	10	16
Black Belt	73	55	13	21
Delta	90	68	13	25
Muscle Shoals	44	26	9	20
Interior Plain	61	32	12	17
Mississippi Bluffs	74	38	10	20
Red River	79	96	14	23

[a]Data for the Arkansas River area not tabulated.
[b]Northern part of Atlantic Coast Plain.
[c]Southern part of Atlantic Coast Plain.
[d]Cotton section.
Source: Woofter, T. J., Jr., "The Subregions of the Southeast", *Social Forces*, October 1934, Vol. 13, pp. 48-49.

In the Red River area, an area with a particularly high rate of tenancy, the median county had only 1 telephone for every 96 inhabitants. On the other hand, in the Muscle Shoals area with the lowest percentage of tenancy, the median county had 1 telephone for every 26 inhabitants. The number of inhabitants

[9]Dickens, Dorothy, *A Study of Food Habits of People in Two Contrasting Areas of Mississippi*, Mississippi Agricultural Experiment Station, Bulletin 245, November 1927.

per passenger car varied less widely than the number per telephone but even so, the number of automobiles tended to be most limited in high tenancy areas. National magazines were rare in all areas but particularly so in the Delta and Red River areas with 90 and 79 percent of tenancy, respectively, as the county medians.

HEALTH

The effects of low income with attendant poor housing and meager diet are evident when measures of health are applied to the cotton tenant household. The lack of screening facilitates the spread of malaria; the primitive water supply and sanitary facilities contribute to typhoid epidemics. The lack of balance in the diet is a major factor in the incidence of pellagra, a disease almost entirely confined to the poor classes in the South. Inadequate food also contributes to digestive disorders.

The close relation between inadequate income and malaria mortality is indicated by a study of the data for 14 southern States[10] in 1933.[11] North Carolina was the only State studied which did not show a significant increase in malaria mortality from 1932 to 1933. The authors of the report, while recognizing the lack of specific evidence, suggest that "the depression, which, in its general impoverishment and degradation of the population, has had no equal since malaria has been recognized as a public health problem in our midst, is probably the most important contributory factor."

Appendix Table 52 and Figure 25 show the incidence of deaths from typhoid and paratyphoid, pellagra, and malaria in the rural sections of the seven southeastern cotton States. These figures do not apply strictly to tenants since they include the whole rural population—owners, tenants, and inhabitants of small towns. Also, since they are based entirely on deaths they do not take into account the great losses due to non-fatal illnesses. They do indicate, however, that the cotton States have the burden of a typhoid and paratyphoid death rate that is twice the national average, and of pellagra and malaria death rates that are more than three times the national average. The incidence of these diseases, as measured by death rates, varies widely within the cotton States.

The high Negro death rate has been attributed largely to ignorance and this is undoubtedly a major factor, but the unhygienic

---

[10] Alabama, Arkansas, Florida, Georgia, Kentucky, Louisiana, Mississippi, Missouri, North Carolina, Oklahoma, South Carolina, Tennessee, Texas, and Virginia.

[11] Faust, E. C. and Diboll, Celeste G., "Malaria Mortality in the Southern United States for the Year 1933", *The Southern Medical Journal*, August 1935, Vol. 28, pp. 757-763.

living conditions, many of which are dictated to the tenant by the system, must also be assigned a major portion of the blame.

FIG. 25

RURAL DEATH RATE FROM SPECIFIC DISEASES
IN SEVEN SOUTHEASTERN COTTON, AND ALL OTHER STATES— 1930

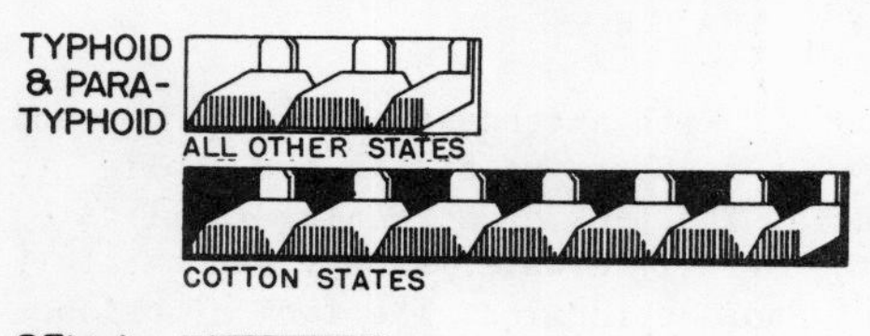

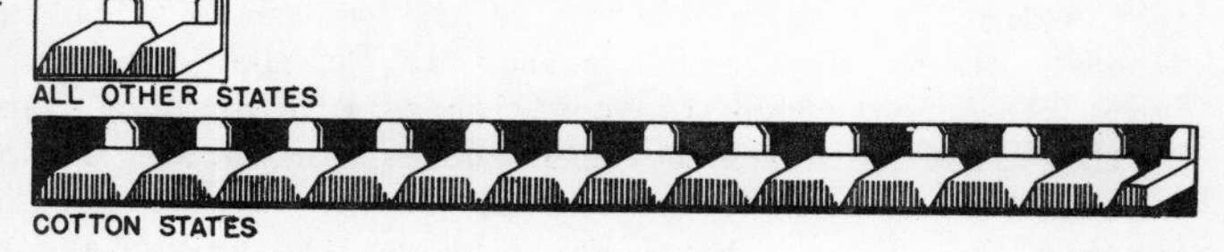

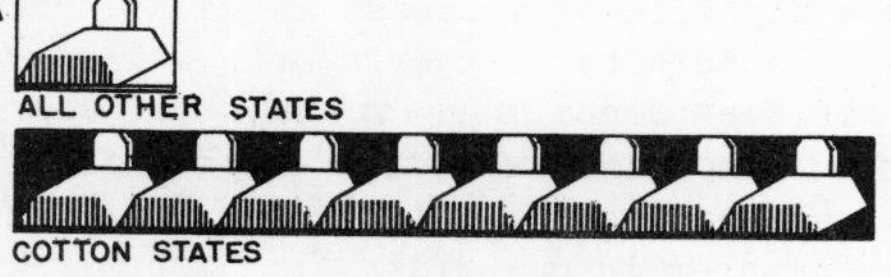

EACH SYMBOL EQUALS ONE DEATH PER 50,000 AF-1565, W.P.A.

SOURCE: MORTALITY STATISTICS 1930, U.S. BUREAU OF THE CENSUS

In the face of such obstacles public health campaigns cannot progress until fundamentals of plantation farming are so altered as to provide a better level of living for tenants.

Chapter VIII

MOBILITY

The tenure system in the Cotton South is characterized by a high rate of mobility among farmers, especially croppers. Every year thousands of cotton tenant farmers place their household goods and other belongings in wagons and trucks and move on to other quarters. Sometimes the destination is another farm at no great distance; sometimes it is a nearby village or town where the tenant expects to engage in industrial activities; and in some instances the move involves a greater distance, as in the case of southern Negroes migrating to northern industrial centers.

The one-crop system, characteristic of a great portion of the area, is an influencing factor. The low educational and occupational standards of the region probably account for some of the "shiftlessness" which is associated with the tenant farmer continually on the move. Low economic standards unquestionably drive some farmers to and fro in search of some way to eke out a more desirable living for themselves and their families.

As a result of these and other factors, three types of movements of farmers and wage hands are relatively frequent and subject to analysis: (1) from one farm to another; (2) from one status to another within agriculture; and (3) from rural areas to town and back.

It must be borne in mind in interpreting the facts in this chapter that the farmers who move to town and remain there do not enter into this study, as the survey was based entirely on reports from families then residing on farms.

MOVEMENT FROM FARM TO FARM

The high rate of mobility from farm to farm among southern farmers, especially tenant farmers, has caused much concern to those who consider stability of residence a condition necessary to successful farming, long-time community relationships the basis for a secure society, and wise use of the land insurance against destruction of the Nation's natural resources. The nature of farming is such that the farmer must spend several years in developing the property he has bought or rented and must give constant attention to the farm if its productivity is to be fully developed and maintained. The farmer who is continually on the move cannot systematically care for fruit trees, shrubbery, and other long-lived vegetation. If he expects to move again soon, he does not find it to his advantage to build fences, to construct drains and terraces, to sow perennial grasses, and to turn under cover crops to conserve and build up the soil. It is difficult for the mobile farmer to provide and care for livestock. Further, it seems that mobility begets more mobility. Because of his apparent instability it becomes increasingly difficult for a mobile farmer to obtain credit facilities to carry out his business operations. Social relations are also continually disrupted as farm families move from community to community.[1]

Whereas excessive mobility of farmers is associated with these disastrous effects, excessive stability may also prove to be harmful to society and to the individual farmer. Lack of mobility of farm populations restricts the outlook to the small horizon of the immediate community and creates resistance to the infiltration of new ideas and behavior patterns. Low mobility may signify an unhealthy indifference to opportunities for individual advancement or it may indicate the existence of a system which impedes free movement.

Between the two extremes lies the more judicious course: the amount of mobility necessary for the successful orientation of the farmer and the amount of stability necessary for him to build a productive farm unit. A certain amount of mobility from farm to farm is a part of the normal life process of the farmer in getting onto the land and becoming initiated into his vocational setting. It seems probable that with farm owners, with tenants who later are to become owners, and in many instances with permanent tenants, one to three moves are often necessary in making the adjustments incident to determining the size and type of farm suited to their needs and capacities.

That mobility among farmers, especially tenants, is related to the changing size of the family has been emphasized in the following words:[2]

> The young tenant farmer and his wife make a few moves in their early life, many shifting from farming to other occupations, then back to farming in an attempt, supposedly, to improve their living conditions. These inter-occupational changes generally involve territorial changes. They occur,

---

[1]Williams, B. O., *Occupational Mobility among Farmers*, Part I, "Mobility Patterns", Agricultural Experiment Station, Clemson, South Carolina, Bulletin 296, 1934, p. 16 ff.

[2]Williams, B. O., *Social Mobility and the Land Tenure Problem*, unpublished manuscript read before Southern Sociological Society, Atlanta, Georgia, April 18, 1936, pp. 5-6.

> generally speaking, in those areas where are found villages and towns having occupations that are freely interchangeable with farming, such as the cotton mills of the industrial regions.
>
> Later on in life, when the children of the tenant family begin to mature and are able to carry on the work of the farm, other moves are made so as to obtain a larger farm. It may be that the tenant farmer will move one, two, or three times in response to the increasing size of his family. At a later period the children begin to leave the parental home and to enter their own occupations. This results in the tenant renting a smaller farm, in response to the decreasing size of his family. After the children have all left home, the farmer and his wife, being old and unable to carry on the physical demands of farm labor, find it necessary to move about as the occasion may demand.

Mobility or stability may be measured in several ways, as indicated in Table 41. The farming history of all families included in the survey that were able to furnish reliable data[3] was recorded.

These indices, the length of residence on the farm on which the family was located in 1934, the total number of farms which the family had occupied since farm life began, and the average length of residence per farm occupied indicate the greater mobility of white tenants than of Negro and greater mobility of wage hands and croppers, with increasing stability as the higher tenure classes are attained. The columns showing age of farmers and average length of time in farming indicate that with the exception of wage hand families these factors are not sufficiently different by tenure status to account for the variations in mobility.

All families had averaged 8.2 years of residence on the farm where they were living in 1934 (Table 41). They had lived on an average of 3.8 farms with an average residence per farm of 5.9 years, thus accounting for their average of 22 years of farming experience. The fact that the residence on the 1934 farm is longer than the residences on previous farms indicates two possibilities, *viz.*, that as the tenants approach maturity they tend to become more stable, and that these plantations,

---

[3]Number of years on present farms was based on 5,049 family schedules: 799 white, 4,250 Negro; 845 wage hands, 2,848 croppers, 708 other share tenants, 648 renters. Other items in the table were based on 4,713 replies: 728 whites, 3,985 Negroes; 777 wage hands, 2,662 croppers, 647 other share tenants, 627 renters.

representing as they do the better cotton lands, are more desirable to the tenants and they stick to these places longer. The obverse of this process is that landlords of the better plantations can pick and choose their tenants and that after a process of trial and error the landlords of the plantations surveyed had built up a tenant personnel with which they were satisfied. In other words, a factor in the stability of a tenant on a plantation is his willingness to stay, modified by the landlord's willingness to keep him.

Table 41—LENGTH OF FARM RESIDENCE OF PLANTATION FAMILIES BY COLOR, BY 1934 TENURE STATUS, BY AGE, AND BY LENGTH OF FARMING EXPERIENCE
(Cotton Plantation Enumeration)

1934 Tenure Status and Color	Average Number of Years Farmed	Average Age of Head[a]	Average Number of Years on 1934 Farm	Average Number of Farms Lived on[b]	Average Number of Years per Farm
Total	22	41	8.2	3.8	5.9
White	21	40	5.9	4.3	4.8
Negro	23	41	8.6	3.7	6.1
Wage hands	15	33	6.5	3.2	4.8
White	12	31	4.3	3.3	3.6
Negro	16	33	6.7	3.2	5.0
Croppers	22	41	6.9	4.0	5.4
White	19	40	4.7	4.4	4.4
Negro	22	41	7.3	4.0	5.6
Other share tenants	28	47	11.2	3.7	7.5
White	24	43	7.2	4.2	5.7
Negro	29	48	12.6	3.5	8.2
Renters	27	47	12.8	3.3	8.2
White	24	44	9.2	4.3	5.7
Negro	28	47	13.7	3.1	9.0

[a] Based on a total of 5,060 families. Data were not available for 111 families.
[b] Since reaching 16 years of age.

When race differences are considered, it is evident that white tenants are more mobile than Negroes. White tenants had lived on the 1934 farm 5.9 years while Negro tenants had lived on the 1934 farm 8.6 years. White tenants had averaged 4.3 different farm residences and Negro tenants 3.7. The average number of years of residence per farm was 4.8 for white tenants and 6.1 for Negroes.

CHANGE IN RESIDENCE BY STATUS

Another trend apparent in the statistics of farm residence in Table 41 is the tendency to stabilize with rise in tenure status. This is to some extent associated with increasing age and with the acquisition of property. The average age of wage hands in the study was 33 years, of croppers 41 years, and of other share tenants and renters 47 years. This difference accounts for the smaller number of farms which wage hands had lived on but is not sufficient to account for other differences in mobility among the tenure classes. The increase in stability is notable in Table 41 with respect both to length of residence on the farm occupied in 1934 and average residence on all farms. Wage hands had lived 6.5 years on the 1934 farm, croppers 6.9 years, other share tenants 11.2 years, and renters 12.8 years. On all farms the average years of residence had been 4.8 for wage hands, 5.4 for croppers, 7.5 for other share tenants, and 8.2 for renters. Facts confirming these trends in mobility were obtained from 1,830 South Carolina farmers in a study made in 1933[4] (Table 42) and from studies of tenants in Alabama[5] in 1933 and in North Carolina in 1934.[6]

The study under immediate consideration did not obtain mobility data for owners. Other studies have shown that the changes in residence of tenants and wage hands have been considerably greater than those of owners. The study of 1,830 South Carolina farmers shows that white owners made an average of 2.9 changes in residence during the period of employment covered, as compared with 5.6 changes made by white tenants. Among Negroes a similar difference appeared, Negro owners making an average of 3.0 changes and Negro tenants making an average of 4.6 changes (Table 42). In other words, on the basis of the number of years employed, white owners had moved about once in every 11 years and white tenants about once in every 4 years, Negro owners every 12 years and Negro tenants every 5 or 6 years. White owners were only 67 percent as mobile and Negro owners only 70 percent as mobile as the entire group, while white tenants were 30 percent more mobile and Negro tenants 7 percent more mobile than the total families studied (Table 42).

The difference in the degree of mobility between tenants and owners, whether white or Negro, is even greater than the above data indicate, owing to the age factor and to the average number of years employed. The age of white owners averaged 54.5 years as compared with 45.5 years for white tenants. The average ages for Negro owners and tenants were 55.0 years and 48.5 years, respectively. Directly related to this is the

---

[4] Conducted by B. O. Williams for the South Carolina Experiment Station, Clemson, South Carolina, in cooperation with the Civil Works, Emergency Relief, and Works Progress Administrations. A 5 percent sample, based on color and tenure of the farmers, was taken in eight counties representing the major soil and type-of-farming areas. Cases were selected by townships in the eight counties on a proportionate basis, according to the 1930 Census. The data were collected by the schedule method, visits being made to each individual farmer. The following counties were represented in the sample: Charleston, Chester, Darlington, Greenville, Hampton, Saluda, Spartanburg, and Williamsburg. All the South Carolina data referred to in this chapter are unpublished material from this study.

[5] Hoffsommer, Harold, *Landlord-Tenant Relations and Relief in Alabama*, F.E.R.A. Research Bulletin Series II, No. 9, November 1935.

[6] Unpublished study of rural families on relief in 1934 in North Carolina conducted by Gordon Blackwell for the North Carolina E.R.A. Sample counties were Alexander, Bertie, Columbus, Greene, Iredell, Onslow, Stokes, Tyrell, and Washington, the mountain region being excluded.

longer average period of employment for owners than for tenants (Table 42).

The fact that white owners and Negro owners had made about the same number of moves is associated with the fact that they were of the same average age and had been employed approximately the same number of years.

Table 42—CHANGES IN RESIDENCE[a] OF 1,830 SOUTH CAROLINA FARMERS, BY NUMBER OF MOVES MADE, BY 1933 TENURE AND BY COLOR

Number of Moves	Number Making Specified Moves						
	Total	White			Negro		
		Total	Owners	Tenants	Total	Owners	Tenants
1	347	207	152	55	140	40	100
2	353	198	141	57	155	58	97
3	260	160	85	75	100	17	83
4	215	109	52	57	106	15	91
5	162	96	27	69	66	6	60
6	134	65	18	47	69	12	57
7	94	57	20	37	37	6	31
8	76	41	7	34	35	3	32
9	60	37	3	34	23	2	21
10	30	15	2	13	15	1	14
11	27	18	5	13	9	1	8
12	15	11	1	10	4	0	4
13	14	5	0	5	9	0	9
14	14	10	1	9	4	0	4
15	7	4	0	4	3	0	3
16	2	1	0	1	1	0	1
17	7	4	1	3	3	0	3
18	3	1	0	1	2	1	1
19	3	2	0	2	1	0	1
20	1	1	0	1	0	0	0
21	3	2	0	2	1	0	1
22	3	2	0	2	1	0	1
Total number of moves made	7,784	4,440	1,481	2,959	3,344	492	2,852
Total number of farmers	1,830	1,046	515	531	784	162	622
Average number of moves	4.3	4.2	2.9	5.6	4.3	3.0	4.6
Index number of moves[b]	100	100	67	130	102	70	107
Average number of years employed	28.8	28.6	32.9	24.4	28.1	34.6	27.5

[a]Every farmer was given credit for one move when he entered employment which generally involved leaving the parental family. In a few instances, the individual remained in the parental family, but he was given credit for one move nevertheless.

[b]The average number of moves made by the 1,830 farmers is taken as 100.

Source: Unpublished data from study conducted in eight representative farming counties by the South Carolina Experiment Station in cooperation with the C.W.A., E.R.A., and W.P.A.

The theory that ownership of farm land acts as a stabilizing influence, tending to bind or tie the farmer to the land, is given further weight by data from the South Carolina study assembled in Tables 43 and 48. They show that whereas white owners had spent 71 percent of their employed years as owners (Table 48), they had made only 39 percent of their total moves in the owner status (Table 43). The 18 percent of their employed years which they had spent as tenants included 39 percent of the moves made. Obviously much of their moving had been done before they entered the owner status. On the other hand white tenants had spent 75 percent of their employed years as tenants (Table 48), and had made 77 percent of their total moves in the tenant class (Table 43). This indicates not only that white tenants tended to remain in the tenant status but also that most of their moving was done as tenants. Moreover, it seems probable that tenants tend to move at a similar rate

Table 43—CHANGES IN RESIDENCE OF 1,830 SOUTH CAROLINA FARMERS, BY OCCUPATION, BY 1933 TENURE, AND BY COLOR

1933 Tenure and Color	Number of Farmers	Number of Moves Made	Percent of Moves Made as					
			Total	Owner	Farm Manager	Tenant	Hired Man	Non-Farming
Owners	677	1,973	100.0	36.8	0.9	39.5	6.4	16.4
White	515	1,481	100.0	39.3	1.2	38.5	3.8	17.2
Negro	162	492	100.0	29.1	-	42.7	14.2	14.0
Tenants	1,153	5,811	100.0	2.1	0.3	79.4	6.9	11.3
White	531	2,959	100.0	3.5	0.7	77.4	3.5	14.9
Negro	622	2,852	100.0	0.7	-	81.4	10.4	7.5

Source: Unpublished data from study conducted in eight representative farming counties by the South Carolina Experiment Station in cooperation with the C.W.A., E.R.A., and W.P.A.

throughout their employed years while owners definitely move less as they grow older.

Comparable data for Negro farmers also indicate that owners did most of their moving before attaining ownership status, while for tenants the proportion of employed years spent as tenants (85 percent) corresponded closely to the proportion of total moves made while in the tenant class (81 percent).

Table 44—NUMBER OF FARMS LIVED ON, NUMBER OF YEARS FARMED, AND NUMBER OF COUNTIES LIVED IN BY NORTH CAROLINA FARM FAMILIES ON RELIEF IN 1934, BY TENURE STATUS

Tenure Status	Number of Families	Average Number of Farms Lived On	Range	Average Number of Years Farmed	Number of Counties Lived In	
					Average	Range
Total	1,515[a]	4.1	1-21	14.2[b]	1.34	1-6
Small farm owners[c]	201	2.7	1-15	18.1	1.17	1-6
Rural home owners[d]	271	2.9	1-15	14.4	1.17	1-4
Tenants and farm laborers	1,043	4.6	1-21	13.5	1.42	1-5

[a]Data not available for 18 of the 1,533 cases enumerated.

[b]Based on data for 1,473 cases.

[c]Owning 10 or more cultivable acres with family of 5 or less members, or 15 or more cultivable acres with family of 6 or more members. This is a functional definition employed in measuring rehabilitation needs and capabilities.

[d]Owning a home but not enough cultivable land to be classified as a small farm owner according to the definition above.

Source: Unpublished data from Study of Rural Relief Families in North Carolina by Gordon Blackwell for the North Carolina E.R.A. Sample counties were Alexander, Bertie, Columbus, Greene, Iredell, Onslow, Stokes, Tyrrell, and Washington, the mountain region being excluded.

Blackwell's study of 1,533 rural relief cases in North Carolina, selected as capable of rural rehabilitation, also shows that tenants are more mobile than owners. Tenants and farm laborers covered in the study had lived on an average of 4.6 farms as compared with 2.9 farms for rural home owners and 2.7 farms for small farm owners (Table 44). The cases were not selected as belonging to the plantation pattern, being distributed over nine counties of North Carolina and representing all types of farming areas of the State except the Appalachian region. Their mobility rates are not directly comparable to those of the plantation families because relief families have been found to have higher mobility rates than non-relief.[7]

The relatively high rate of mobility among tenant farmers, as compared with owners, in the Cotton South is undoubtedly tied up with the system of land renting. The nature of the one-crop system and the credit structure have already been described (chapters IV and V). The lack of written contract between tenant and landowner, and the fact that the tenant has no legal claim and receives no recompense for improvements he may make on the property, deprive the cotton tenant farmer of that incentive which leads to stability. But even though the tenant remains

Table 45—DISTANCE OF MOVES MADE BY 1,830 SOUTH CAROLINA FARMERS, BY COLOR, AND BY 1933 TENURE

Color and Tenure	Number of Farmers	Number and Proportion Moving Specified Distance											
		Total Moves[a]		Within County[b]		Adjoining County		Other South Carolina County		Adjoining State		Other State	
		Number	Percent	Number	Percent	Number	Percent	Number	Percent	Number	Percent	Number	Percent
Total	1,830	5,954	100.0	5,028	84.4	516	8.7	144	2.4	154	2.6	112	1.9
White													
Owner	515	966	100.0	823	85.2	57	5.9	34	3.5	23	2.4	29	3.0
Tenant	531	2,428	100.0	1,946	80.2	280	11.5	65	2.7	92	3.8	45	1.8
Negro													
Owner	162	330	100.0	289	87.6	7	2.1	10	3.0	8	2.4	16	4.9
Tenant	622	2,230	100.0	1,970	88.3	172	7.7	35	1.6	31	1.4	22	1.0

[a]Excludes one entrance move for each farmer.
[b]County in which the farmer was living at time the survey was made (1933).
Source: Unpublished data from study conducted in eight representative farming counties by the South Carolina Experiment Station in cooperation with the C.W.A., E.R.A., and W.P.A.

relatively fixed, he is handicapped so far as developing his property is concerned, for his agreement with the landlord often requires him to devote a high proportion of his land to cotton.

If the tenant is to be encouraged to remain on one farm and to conserve and develop the soil, his lease should be drawn up in such a way as to reward him for making permanent and durable improvements on the land and buildings and other non-movable fixtures about the premises. Perhaps a flexible lease would be desirable and might be drawn up in terms of 1, 3, 5, or 10 years. Certainly the lease should state explicitly the basis for dividing and allocating the rewards for permanent and durable improvements; it should prescribe the conditions upon which rent is to be paid; and it should be entered into as between the lessor and the lessee with a clear and full understanding of the rights and responsibilities of each.

---

[7]McCormick, T. C., *Comparative Study of Rural Relief and Non-Relief Households*, W.P.A. Research Monograph II, pp. 17-18, 1935.

DISTANCE OF MOVES

In spite of their frequent moving, few of the South Carolina farmers had moved great distances. Out of a total of 5,954 moves made by the 1,830 farmers, 84.4 percent were made within the county in which the farmers were living in 1933 (Table 45). The proportions of moves made within the county are fairly consistent for the different tenure groups. However, the highest percentage was for Negro tenants, 88.3 percent of their moves having been made within the county. Similarly, Negro owners had made 87.6 percent of their moves within the county. The proportion of moves made within the county was lowest for white tenants, 80.2 percent. White owners had made 85.2 percent of their moves within the county.

Although the white tenants had made the highest proportion of moves outside the home county of any of the tenure group (19.8 percent), most of these moves were made to an adjoining county (11.5 percent).

Blackwell's study in North Carolina shows similar results. Few of these farm families had moved great distances. While these were relief families believed capable of rehabilitation, yet there is reason to believe that they were somewhat representative of their respective groups. Table 44 indicates that these farmers and farm laborers had lived on farms in less than one and one-half counties, on the average, during their years of employment.

CHANGE OF STATUS

Movements within the farming occupation can be either up or down the agricultural ladder, upward as defined in this survey meaning the gradations from wage hand to share-cropper, to share tenant, to renter, to owner. Each of these grades generally means some advance over the one before it, in income, in working conditions, or in property status. Hence movement up this ladder usually spells a certain degree of progress for the farmer, while movement down the ladder means that he is probably losing ground. Rapid progress from share tenant to farm owner may indicate, however, merely that the erstwhile tenant has saddled himself with debt, and that with hard times he may return to his former status.

Number of Years in 1934 Tenure Status

In tracing the progress of plantation families up or down the agricultural ladder the average time that the families had remained continuously in their 1934 tenure status was first determined (Table 46). Taking all groups, the families had been engaged continuously for an average of 12.5 years in the status they occupied in 1934. Renters and other share tenants, the groups with the highest average number of years in agriculture (Table 41), had been in their 1934 tenure status for the longest periods of time, followed by share-croppers and wage hands. In all groups, the average number of continuous years in the 1934 status accounted for at least half of the total number of years employed in agriculture.

The number of heads of plantation families who had farmed in tenures or classes other than those occupied in 1934 and the number of years spent in each previous status are shown in Table 47.

Table 46—NUMBER OF CONTINUOUS YEARS SPENT IN 1934 TENURE STATUS BY PLANTATION FAMILIES, BY 1934 TENURE STATUS, AND BY COLOR
(Cotton Plantation Enumeration)

1934 Tenure Status and Color	Total Families Reporting[a]	Average Number of Continuous Years in 1934 Tenure Status
Total	5,001	12.5
White	788	10.1
Negro	4,213	13.0
Wage hands	828	8.1
White	82	6.4
Negro	746	8.3
Croppers	2,816	13.1
White	405	9.9
Negro	2,411	13.6
Other share tenants	712	13.8
White	180	11.7
Negro	532	14.5
Renters	645	14.5
White	121	11.2
Negro	524	15.2

[a]Data not available for 170 families.

Almost two-thirds (63 percent) of the 1934 cropper families had at some time been wage hands and had worked in that status an average of 6 years; 22 percent had worked as other share tenants, for an average of 9 years; 18 percent had been renters for an average of 8 years; and less than 3 percent of the croppers had been farm owners for an average of 11 years (Table 47). The trend in mobility was therefore up the ladder, 63 percent of the share-croppers having come up from the status of wage hands, as against 43 percent moving down from other tenures. The fact that almost three-fourths of all plantation families were share-croppers or wage hands seems to indicate the difficulty of ascending the agricultural ladder under the plantation system.

Half of the other share tenants had risen from the status of wage hands and three-fifths had formerly been croppers. One-fourth had dropped from the status of renter and about 4 percent from the status of owner.

Table 47—NUMBER OF YEARS FARMED IN EACH TENURE STATUS BY HEADS OF PLANTATION FAMILIES, BY 1934 TENURE STATUS AND BY COLOR
(Cotton Plantation Enumeration)

1934 Tenure Status	All Heads of Families		Previous Tenure Status														
			Wage Hands			Croppers			Other Share Tenants			Renters			Owners		
	Number	Average Number of Years Farmed	Number	Per-cent	Average Number of Years Farmed	Number	Per-cent	Average Number of Years Farmed	Number	Per-cent	Average Number of Years Farmed	Number	Per-cent	Average Number of Years Farmed	Number	Per-cent	Average Number of Years Farmed
Total	5,085	22	3,456	68.0	7	3,929	77.3	13	1,533	30.1	11	1,406	27.6	12	134	2.6	11
White	813	20	488	60.0	6	582	71.6	10	315	38.7	11	270	33.2	10	82	10.1	11
Negro	4,272	22	2,968	69.5	7	3,347	78.4	13	1,218	28.5	12	1,136	26.6	13	52	1.2	10
Wage hands	859	15	859	100.0	9	352	41.0	9	65	7.6	8	62	7.2	9	7	0.8	8
White	87	11	87	100.0	7	26	29.9	7	11	12.6	7	5	5.7	7	4	4.6	6
Negro	772	15	772	100.0	10	326	42.2	10	54	7.0	8	57	7.4	9	3	0.4	11
Croppers	2,866	22	1,793	62.6	6	2,866	100.0	14	636	22.2	9	515	18.0	8	73	2.5	11
White	420	19	245	58.3	6	420	100.0	11	95	22.6	9	88	21.0	8	41	9.8	11
Negro	2,446	22	1,548	63.3	6	2,446	100.0	15	541	22.1	9	427	17.5	8	32	1.3	10
Other share tenants	712	28	363	51.0	6	433	60.8	11	712	100.0	14	181	25.4	11	28	3.9	10
White	184	24	89	48.4	6	91	49.5	9	184	100.0	13	55	29.9	8	19	10.3	10
Negro	528	29	274	51.9	6	342	64.8	12	528	100.0	15	126	23.9	12	9	1.7	8
Renters	648	27	441	68.1	7	278	42.9	9	120	18.5	10	648	100.0	17	26	4.0	13
White	122	24	67	54.9	6	45	36.9	10	25	20.5	10	122	100.0	13	18	14.8	14
Negro	526	28	374	71.1	7	233	44.3	9	95	18.1	10	526	100.0	17	8	1.5	9

A predominantly upward mobility trend is evident for the renters. More than two-thirds once had been wage hands, 43 percent had been share-croppers, and 19 percent had been other share tenants. Only 4 percent had dropped from the status of owner.

On the other hand, while many of the wage hands had never held any other status, 41 percent had been share-croppers, 8 percent formerly had been other share tenants, and 7 percent had been renters. Almost none (0.8 percent) had occupied the status of owner.

It is evident that the plantation families had been mobile within agriculture as regards tenure status. At least half of all the family heads in the different groups above the wage hand status had once been wage hands. More than 40 percent of the farmers in each status had spent 9 years or longer, on the average as share-croppers. A small proportion of the farmers in each group once had been owners for an average of 10 years or more.

Table 48—CHANGES IN STATUS OF 1,830 SOUTH CAROLINA FARMERS, BY 1933 TENURE, AND BY COLOR

1933 Tenure and Color	Number of Farmers	Years Employed	Percent of Employed Years Spent as					
			Total	Owner	Farm Manager	Tenant	Hired Man	Non-Farming
Owners	677	22,238	100.0	68.7	0.4	19.8	3.4	7.7
White	515	16,695	100.0	71.3	0.5	18.0	1.3	8.9
Negro	162	5,543	100.0	62.4	-	25.3	8.1	4.2
Tenants	1,153	29,686	100.0	4.3	0.5	80.4	5.5	9.3
White	531	12,753	100.0	8.0	1.0	74.7	3.0	13.3
Negro	622	16,933	100.0	1.5	0.1	84.8	7.4	6.2

Source: Unpublished data from study conducted in eight representative farming counties by the South Carolina Experiment Station in cooperation with the C.W.A., E.R.A., and W.P.A.

In all groups, larger proportions of Negroes than of whites had been wage hands and share-croppers at sometime. Smaller proportions of Negroes than of whites had moved down the agricultural ladder because their status tended to be lower in general (Table 47).

The study of 1,830 South Carolina farmers in 1933 gives additional data on changes in status, including information about owners. Owner-operators interviewed had spent slightly more than two-thirds of their employed lives as owners (Table 48). On the average, they had been tenants about one-fifth of their employed years, and had worked in non-farming occupations about one-twelfth of their working lives. Very little of their time had been spent as wage hands or farm managers.

South Carolina tenant farmers had spent a much greater part of their employed years in their 1933 status than had owners, having spent about four-fifths of the years as tenants and 4.3 percent of their employed years in the status of owner, indicating slight movement down the ladder. Negro tenants had been somewhat more stable occupationally than whites, spending 85 percent of their employed years as tenants, compared with 75 percent for whites.

The tendency to move up the agricultural ladder is undoubtedly accelerated in periods of prosperity. Conversely, in years of unprofitable operations there is a tendency toward shifting downward. During the depression years 1929-1932 many farm owners lost their places, numerous renters and share tenants lost their work stock and equipment, and thousands of croppers became casual farm laborers. With higher cotton prices improvement of status became possible for many families.

Such change in status is indicated by a study of 1,703 rural families in North Carolina,[8] made by Hamilton in 1934-1935 after incomes had been improved by the A.A.A. He found that out of 185 farm laborers in 1934, 43 had shifted up the ladder in 1935 into the cropper, renter, and owner groups. Of 400 croppers, 22 shifted up the ladder as contrasted with 19 who became farm laborers. Of 356 renters, 8 moved up the ladder and 19 dropped to the status of cropper or laborer. Of 483 owners in 1934, 9 shifted down the ladder in 1935.

In comparison, only 21 of the 202 farm laborers in 1931 shifted up the ladder in 1932, only 16 of the 380 croppers, and 4 of the 321 renters. Of the 472 owners in 1931, 12 became renters or croppers in 1932.[9]

Tenure and Occupational Mobility

Analysis of the occupational stability (Appendix Table 53) of South Carolina farmers in 1933 shows that 27 percent of the white owners had always been owners, never having experienced a change in tenure status. This was also true of 14 percent of the Negro owners, who began as owners and remained in that status. Similarly, 45 percent of the white tenant farmers and 52 percent of the Negro tenants had always been tenants throughout their employed lives. Approximately 35 percent of the white

---

[8]Hamilton, Horace C., *The Relation of the Agricultural Adjustment Program to Rural Relief Needs in North Carolina*, North Carolina Experiment Station, November 1935 (mimeographed).
The survey included all households in selected townships or sections of townships in Johnston, Robeson, Richmond, Rutherford, and Caswell Counties. The optimistic note sounded in this report perhaps should be qualified. Displacement of croppers had not been as extensive in these sample counties as in other cotton and tobacco counties in eastern North Carolina. Data from the 1930 and 1935 Censuses of Agriculture reveal that, when combined, these five sample counties had a 1930-1935 decrease of only 2.9 percent in the number of croppers, while all cotton and tobacco counties in the State had a decrease of 7.7 percent (Appendix Table 60). On the other hand, the five sample counties had an increase between 1930 and 1935 of 9.5 percent in the number of "other tenants" (share tenants and renters) while all cotton and tobacco counties in the State had an increase of only 4.8 percent.

[9]*Idem*, p. 2.

owners started out as tenants and moved directly from that status to ownership, remaining owners throughout the subsequent years of their employment until the date of the study (1933). Approximately 11 percent of the white owners began as non-farmers and transferred directly to farm ownership, remaining subsequently in that category.

The evidence presented in Appendix Table 53 shows that white farmers had traveled three main roads to ownership: (1) direct to ownership when they began their occupational career; (2) from tenancy to ownership; and (3) from non-farming to ownership. The Negroes had traveled a somewhat different road. A slightly

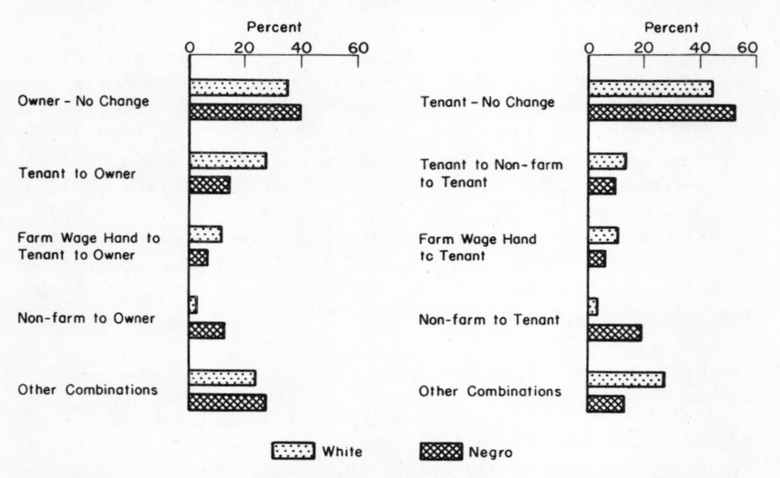

FIG. 26 - OCCUPATIONAL HISTORY OF 1,830 SOUTH CAROLINA FARMERS, BY 1933 TENURE AND BY COLOR

Source: Study conducted by South Carolina Experiment Station in cooperation with the C.W.A., E.R.A., and W.P.A.

AF - 1459, W.P.A.

higher proportion, 40 percent, began as tenants and went directly to ownership; 14 percent began as owners and remained owners; and 12 percent began as hired men, then became tenants, and finally owners. The patterns of tenure-occupation combinations are presented with greater clarity in Figure 26.

The net long-time movements were upward until 1910—as indicated by the increase in owners and tenants at the expense of laborers. However, since 1910 the number and proportion of owners have decreased and the number and proportion of tenants increased.[10] Croppers were enumerated separately for the first

---

[10] See chapter I, Table 3.

time in 1925. Since that date they have increased in number and proportion of the total, indicating a net downward movement into this class.

RURAL-URBAN MOBILITY

Movements from agriculture to non-agricultural industries and *vice versa* are usually, but not always, linked with residential mobility. A farmer leaves his farm to take a job in a factory town, or an unemployed urban worker goes back to the farm. As in the case of movement from farm to farm, a certain amount of inter-occupational mobility is necessary for making

Table 49—MOVES TO TOWN BY PLANTATION FAMILIES, BY 1934 TENURE STATUS, AND BY COLOR
(Cotton Plantation Enumeration)

1934 Tenure Status and Color	Total Families Reporting[a]	Percent Making Specified Number of Moves to Town					
		Total	0	1	2	3	4 and Over
Total	4,838	100.0	87.5	10.1	1.4	0.5	0.5
White	728	100.0	83.1	13.4	1.9	0.6	1.0
Negro	4,110	100.0	88.3	9.5	1.3	0.5	0.4
Renters	628	100.0	86.1	12.1	1.0	0.3	0.5
White	107	100.0	78.5	19.7	0.9	–	0.9
Negro	521	100.0	87.7	10.6	1.0	0.4	0.3
Other share tenants	672	100.0	92.1	6.4	0.7	0.3	0.5
White	167	100.0	85.6	10.8	1.8	0.6	1.2
Negro	505	100.0	94.3	4.9	0.4	0.2	0.2
Croppers	2,715	100.0	86.5	11.1	1.4	0.7	0.3
White	372	100.0	81.7	14.2	2.2	0.8	1.1
Negro	2,343	100.0	87.3	10.6	1.2	0.6	0.3
Wage hands	823	100.0	88.1	8.4	2.6	0.5	0.4
White	82	100.0	90.3	7.3	2.4	–	–
Negro	741	100.0	87.9	8.5	2.6	0.5	0.5

[a]Data not available for 333 families.

vocational adjustments. Broad general movements back to the farm or away from the farm, however, signalize vital changes in the economic situation, and especially in the condition of industrial activity. When industry booms, workers leave the farm for the cities; when industry slumps, workers return to the farm.

The plantation families included in this study had remained closely tied to the soil during their years of employment, as measured by the proportion that had moved to town and back. Only a small proportion had moved and these had moved only a few times (Table 49). For all cases for which data were available (4,838), only 1 in 8 had made 1 or more moves to town and only 1 in 40 had moved more than once. The proportion of Negroes that had never moved to town was 88.3 percent as compared with 83.1 percent for the whites.

Comparing the different tenure groups, it was found that other share tenants made fewer moves to town than any other group, while renters and croppers made the most. Among whites the most moves to town had been made by renters, followed by share-croppers, other share tenants, and wage hands, the last group having made the fewest moves of all. Among Negroes the most moves to town were made by share-croppers, followed by renters, wage hands, and other share tenants.

The fact that only a few of the plantation families in all classes had moved to town more than once makes it evident that those resident on the plantations studied had very little inter-occupational mobility but had remained to a great extent in agriculture.

The South Carolina study further showed that white tenants had spent only 13.3 percent of their employed lives in non-farming and Negro tenants had spent only 6.2 percent of their employed lives in non-farming (Appendix Table 53). Comparable proportions for owners were 8.9 percent and 4.2 percent, respectively. These differences by color are probably due in large part to the lack of opportunities outside of agriculture for Negroes.

These plantation families, it would seem, grow up and perpetuate the culture of the farming occupation. They remain to a great extent in constant touch with farming, and there is little inter-occupational mobility. Thus, the attitudes of the people are highly conditioned by agricultural habits, and they tend to reflect, accordingly, the behavior patterns characteristic of the local agricultural groups. With low inter-occupational mobility, or with few of the families moving back and forth to town, there is little opportunity for the spread of ideas from the outside world into the plantation system.

Plantation families in this study, and farm families for whom data were obtained in other studies, were farming at the time they were daily enumerated. Hence, so far as inter-occupational mobility is concerned, they include only those who had moved from farming to non-farming and back again or those who started out in non-farming occupations and shifted to farming. These form only a very small proportion of the total number who moved to town.

It was noted in chapter I that nearly 4 million persons born in the rural Southeast were living in other sections in 1930. It is impossible to tell how many of these moved directly from farming to urban industries and how many were below working age when they moved. However, the age distribution of Negroes in Michigan in 1930 (Appendix Table 54), most of whom were recent migrants from southern States, indicates by the concentration in the young adult age groups that most of these left the South just before or soon after they reached the age for entering agriculture on their own account.

Hamilton's study of migration by age groups from southern rural town populations indicates a similar concentration of white migrants in the early adult age groups.[11] After these migrants become accustomed to city life few ever return to farms.

MOBILITY BY COLOR

Throughout this chapter various differences have been noted regarding the mobility of Negroes and whites. Negroes on plantations appear to be less mobile in most respects than whites. In view of the fact that Negro families outnumbered the white families on the plantations covered by this survey at the rate of more than five to one, the special character of Negro mobility should be noted.

For instance, Negro families had lived 8.6 years on the farm where they resided in 1934, while the whites had lived there an average of only 5.9 years (Table 41). For the large group of share-croppers the figures were 7.3 years for Negroes, as compared with 4.7 years for whites. Similar comparisons could be made for all tenures.

Negro heads had lived on each farm, since the worker was 16 years of age, an average of 6.1 years, while the comparable figure for whites was only 4.8 years per farm (Table 41). Negro share-croppers had lived 5.6 years on each farm and white share-croppers 4.4 years.

Negro families had also lived on fewer farms than had whites, the white plantation families having lived on an average of 4.3 farms as compared with 3.7 farms for the Negro families (Table 41). White croppers had lived on an average of 4.4 farms as compared with 4.0 for Negro croppers.

These differences are even greater than the data cited indicate because Negro farmers are older and have spent more years in farming than have comparable white groups.

The fact that the Negroes were less mobile than the whites should not necessarily be interpreted to mean that they were more successful farmers as a result of their relative stability, although it seems true that most landlords prefer good Negro tenants to white tenants. This stability or relative immobility may be of that type referred to above which arises out of the existence of a system which limits personal and individual initiative in making choices of residence. The relative stability of the Negro families may indicate that Negroes are less free to circulate territorially than whites and that their stability is the result of conditions to some extent forced upon them by circumstance. The Negro is certainly in a less favorable bargaining position than the white.

---

[11] Unpublished C.W.A. study of migrants from seven southern States.

This indicates that mobility, which gives fluidity to a population, may work one way for one social group, and another way for another group. The conditions incident to mobility, or stability, are important and gauge the efficiency or inefficiency of mobility as a part of the social mechanism.

Comparisons similar to those made above may also be made on the basis of data from the South Carolina study. White farmers undoubtedly have access to more industries than Negro farmers in South Carolina and therefore have a greater opportunity for inter-occupational mobility. This is linked with the relative differences regarding mobility from farm to farm mentioned above. The fact that access to industry is more open to white farmers makes the number of territorial moves greater for white farmers, since a change from farming to non-agricultural industry, or *vice versa*, is generally accompanied by a change in residence.

Of the Negro plantation families in this study, 88.3 percent had never moved to town (presumably to take a non-agricultural occupation), as compared with 83.1 for the whites (Table 49). The South Carolina study showed (Table 48) that white tenant farmers had spent 13.3 percent of their employed lives in non-farming as compared with only 6.2 percent for the Negroes.

Negroes were also less mobile within agriculture. Heads of families had remained in their 1934 tenure status for an average of 13.0 years, as compared with 10.1 years for the whites (Table 46) and the same relationship holds for each status. The Negro croppers in this plantation study had remained in their 1934 tenure status an average of almost 14 years, as compared with an average of 10 years for the white croppers. Negroes who were other share tenants had worked in this tenure continuously for an average of 15 years, compared with 12 years for white tenants. The Negro renters had farmed an average of 15 years in the renter status, and the white renters an average of 11 years. The Negro wage hands had worked an average of 8 years, and the white wage hands an average of 6 years in that status.

The study of South Carolina farmers revealed that Negro tenant farmers had moved on the average 4.6 times during their employed life and the white tenant farmers 5.6 times (Table 42). White owners and Negro owners, however, had done about the same amount of moving during their employed lives.

Chapter IX

EDUCATION

The plantation system as a step in the evolution from slavery to free labor was a device for the cultivation of large tracts of land with relatively ignorant labor. It is essentially a system whereby the landlord furnishes the capital and brains and the tenant only the brawn. The low level of intelligence of the tenants has tended to perpetuate the system.

Many of the problems which aggravate the evils of the plantation system are capable of alleviation by education. Failure to use the land intelligently, to diversify crops, to accumulate and improve domestic animals, to strive more actively for a higher standard of diet, housing, and health, all rest in part on the lack of education.

For the present generation of tenants schooling is a thing of the past. They can be reached only by adult training programs such as the farm and home demonstration extension activities, and they cannot be reached by these programs unless the landlord is cooperative, since the influence of practically all outside programs on the plantation channels through the landlord who has almost absolute authority over what shall and what shall not take place on his land.

The fact that many of the tenant's ways of living are not only habits of a lifetime but matters of tradition means that they can only be changed slowly, that quick panaceas will not succeed, and that change will await the education of a new generation to different ways of living.

Although the southeastern cotton States have shown substantial advances in education during the past 25 years, they are still trailing all other sections of the country in the schooling made available to their children (Figure 27). They are at the bottom of the list when rated by conventional educational indices—literacy, school attendance, average grades completed, value of buildings, length of school term, salaries paid to teachers, and funds expended per pupil.

Added to a background of deficiencies in the educational system over a period of generations are regional characteristics which inevitably place the Southeast in a poor light when compared with other sections. The population of the Southeast is predominantly rural, and rural schools throughout the country have lower standards than urban schools. Furthermore, owing to a high birth rate and a high rate of migration of adults to other parts of the country prior to the depression, there is a disproportionate number of children of school age in the Southeast in relation to the number of productive adults and the value of taxable property. The practice of race segregation further complicates the problem of adequate schooling. The agricultural system of the Southeast interferes with school attendance since it requires the labor of children during the school term and causes families to make frequent moves which break into the school year. Along with all these handicaps goes the general apathy toward public education, which extends to those who need education most, as well as to those in a position to provide it.

The educational and cultural level of the southeastern population is of considerable national as well as sectional concern. It has been estimated that about 12 percent of the children born in Alabama or Mississippi make their life contribution in some other part of the United States and migration has had a similar result in the other States of the Southeast.[1]

The persons most directly affected by the low educational standards of the Southeast are the tenant farmers, particularly the share-croppers. The Negro share-cropper is affected most seriously of all. The excuse is frequently given that the discrepancy between the level of education in the South and that in other sections of the country is accounted for by the large Negro element in the population. Although such a statement may be true, it has no value in solving the practical problems at hand. The fact that 84 percent of the total tenant households covered in the present survey were Negro emphasizes forcefully that the Negro is basic to the plantation tenant system and cannot be ignored in the consideration of its problems.

ILLITERACY

The Southeast still leads the country in illiteracy. The ratio of illiterates to total population 10 years of age and over ranged from nearly 15 percent in South Carolina to less than 1 percent in Iowa in 1930 (Appendix Table 55 and Figure 28). Of the seven States reporting 10 percent or more of illiteracy, five fell within the southeastern cotton region. Even more marked was the excessive ratio of illiterates 21 years of age and over, South Carolina again ranking highest with more than 18 percent, Mississippi and Louisiana following closely in line.

---

[1]Woofter, T. J., Jr., "Southern Population and Social Planning", *Social Forces*, October 1935, Vol. XIV, p. 19.

FIG. 27 - RELATIVE RANKING OF THE STATES IN EDUCATION, 1930

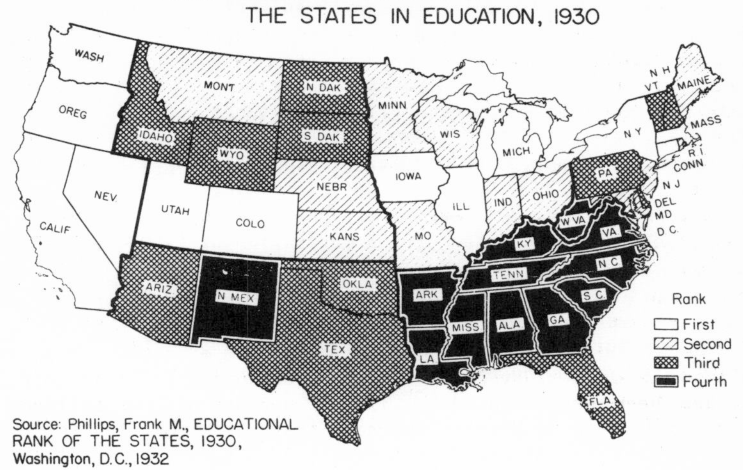

Source: Phillips, Frank M., EDUCATIONAL RANK OF THE STATES, 1930, Washington, D.C., 1932

AF-2029, W.P.A.

FIG. 28 - ILLITERACY IN THE POPULATION 10 YEARS OLD AND OVER, 1930

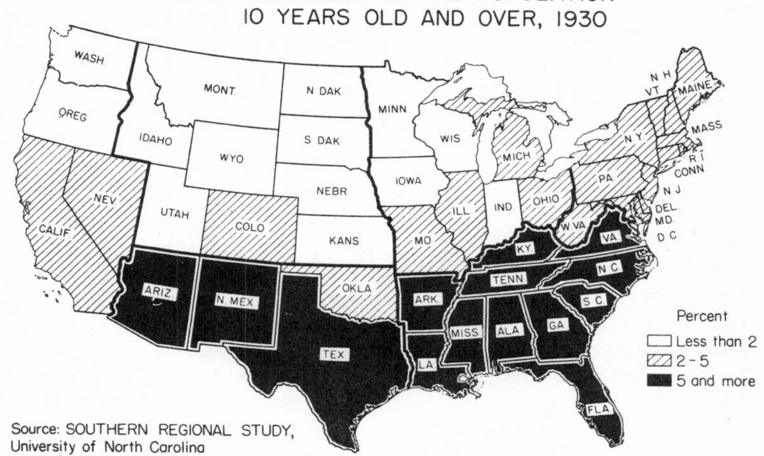

Source: SOUTHERN REGIONAL STUDY, University of North Carolina

AF-2031, W.P.A.

For rural areas alone the proportions were even higher. In the seven southeastern States the range in percent of illiteracy for the rural population 10 years of age and over was from 7.8 percent in Arkansas to 18.3 percent in Louisiana. The proportions for those 21 years of age and over ranged from 10.1 percent in Arkansas to 23.7 percent in Louisiana. Five of the seven States reported 15 percent or more of the rural adults as illiterate in 1930.[2]

Since there are very few foreign-born persons in the southeastern cotton States, the Census definition of literacy means largely the ability to read and write English. Although Census enumerators were instructed not to return a person as literate simply because he could write his name, obviously a great many of those returned as literate were actually unqualified in the essentials of reading and writing. An intensive study in Alabama,[3] covering more than 1,000 farm families receiving relief, showed that approximately one-third of the adults were essentially illiterate[4] and an additional one-third were barely literate, having had the advantage of only a fourth, fifth, or sixth grade education. The fourth grade represented the modal school attainment for the entire group. Among the Negroes more than half of the adults were essentially illiterate. Obviously, this Alabama sample should not be considered as representative of all cotton farmers since it included only those who were on the relief rolls in December 1933. Yet the existence of such a relatively large group is sufficiently alarming to merit serious attention.

Not all illiterates in the Southeast were left over from the past generation. Besides high rates of illiteracy in the general population 21 years of age and over, all the cotton States showed more than 3 percent illiteracy among rural children 10 to 20 years of age in 1930—evidence of the persisting inadequate educational facilities of these States.[5]

ENROLLMENT, ATTENDANCE, AND GRADES COMPLETED

To judge the educational standards of the South by the ratio of total population enrolled in public schools would be misleading. By this index all of the seven southeastern cotton

---

[2] *Fifteenth Census of the United States: 1930*, Population Vol. III, Table 7.

[3] Hoffsommer, Harold, *Education and Rehabilitation in Alabama Farm Households Receiving Relief*, Alabama Polytechnic Institute Bulletin, Vol. XXX, No. 7, July 1935.

[4] Included in this category were those who had either never been to school or those who had not passed beyond the third grade.

[5] By States the percentages were as follows: Alabama, 6.9 percent; Arkansas, 3.2 percent; Georgia, 5.8 percent; Louisiana, 8.0 percent; Mississippi, 5.9 percent; North Carolina, 4.8 percent; and South Carolina, 9.7 percent. *Fifteenth Census of the United States: 1930*, Population Vol. III, Table 7.

States except Louisiana showed in 1931-1932 a larger ratio of total population enrolled in public schools than the average for the United States as a whole. Mississippi led with 28.6 percent, as compared with 21.1 percent for the United States.[6]

FIG. 29-PERCENT OF POPULATION UNDER 20 YEARS OF AGE, 1930

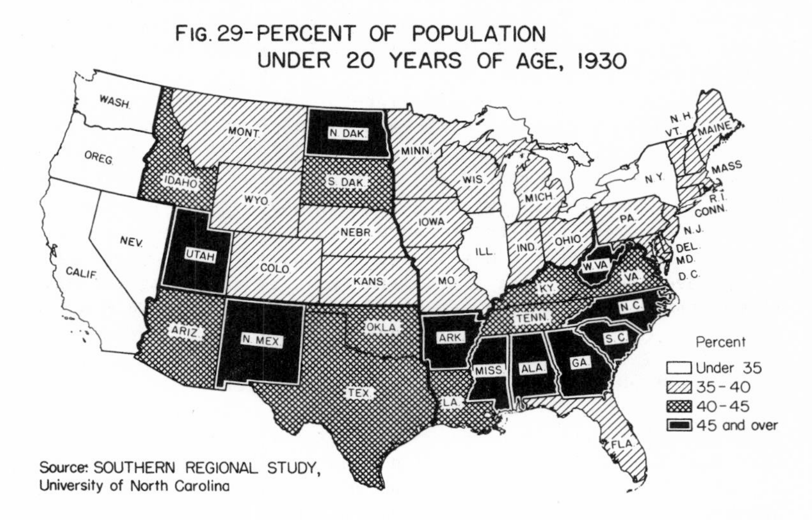

Source: SOUTHERN REGIONAL STUDY,
University of North Carolina

AF-2033, W.P.A.

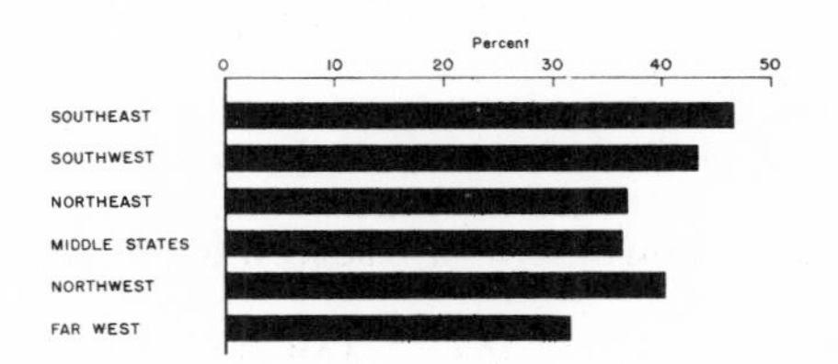

PERCENT OF POPULATION UNDER 20 YEARS OF AGE, 1930

Source: SOUTHERN REGIONAL STUDY,
University of North Carolina

AF-2035, W.P.A.

This situation in the Southeast is merely a reflection of the relatively large percentage of persons of school age in this region (Figure 29). The implication for education is that these States, with low per capita taxable wealth, have disproportionately large numbers of children to be educated.

---

[6] *Biennial Survey of Education, 1930-1932,* "Statistics of State School Systems, 1931-1932", U. S. Department of the Interior, pp. 49-50.

A more significant index is the average number of days attended by each pupil enrolled. In this respect the southeastern States have the lowest rating of any in the country. Of the six States showing an average of less than 125 days attendance at school per enrolled pupil, all except one (Kentucky) were within the southeastern cotton region. Mississippi stood at the bottom of the list with an average of 98.1 days, as contrasted with Illinois which ranked first with 163.2 days.[7]

For rural pupils alone the average number of days of school attendance in 1931-1932 was even less, ranging from 90 in Georgia to 122.5 in North Carolina,[8] compared with 132.4 for the

Table 50—EDUCATION OF CHILDREN OF RURAL RELIEF AND NON-RELIEF HOUSEHOLDS IN THE OLD SOUTH COTTON AREA, BY COLOR, OCTOBER 1933

Area	Percent of Children 5-25 Years of Age Still in School		Percent of Children 12-19 Years of Age Who Completed Grade School		Percent of Children 15-23 Years of Age Who Completed High School	
	Relief	Non-relief	Relief	Non-relief	Relief	Non-relief
All Areas (47 counties)	68	68	47	61	11	27
Old South Cotton Area (5 counties)	51	58	11	26	4	8
White	54	59	17	44	7	14
Negro	49	57	6	9	1	3

Source: McCormick, T. C., *Comparative Study of Rural Relief and Non-relief Households*, W.P.A. Research Monograph II, 1935, pp. 92-93

rural United States as a whole, and 166.8 days for Illinois. Of six States with a rural attendance of less than 110 days, all except Kentucky were within the southeastern cotton region.

That the Negro schools compare unfavorably with the white is common knowledge. The average number of days attended by each Negro child enrolled in both urban and rural schools in 1931-1932 was less than that attended by white children in each of the seven southeastern cotton States. The difference ranged from 23 days in Alabama and North Carolina to 51 days in South Carolina.[9]

The cotton States also led in proportion of persons who had never attended school. A recent survey of rural relief and non-relief households covering 47 counties representing 13 major agricultural areas of the United States revealed that in the 5 counties in Alabama, Arkansas, and North Carolina representing the Old South Cotton area, 16 percent of all heads of non-relief households (in 1933) had never attended school. Among the Negroes this proportion reached 26 percent as compared with only 7 percent for the whites (Table 50). Among the relief households

---

[7] *Biennial Survey of Education, 1930-1932, op. cit.*, p. 60.

[8] *Idem*, p. 104.

[9] *Idem*, p. 95.

more than a third of the Negro heads and a fifth of the white heads had never attended school.[10]

The educational differences between the open country population of the Southeast and of the United States indicate that the average grade attainment of the Southeast is about a grade and a half behind the United States total, and that the average grade attainment of the Negroes alone is three and a half years behind. Although these comparisons (Table 51) are made on the basis of the relief population, much in the same sectional differential holds in the non-relief group.[11] The grade attainments of the generation in school or just out of school are much greater than those of the older generation, the 18 to 20 year group in the Eastern Cotton Area being a full grade ahead of

Table 51—MEDIAN GRADE IN SCHOOL COMPLETED BY HEADS AND MEMBERS 10-64 YEARS OF AGE OF OPEN COUNTRY RELIEF HOUSEHOLDS IN NINE REPRESENTATIVE AGRICULTURAL AREAS AND IN THE EASTERN COTTON AREA, BY AGE GROUPS, OCTOBER 1935

Age	Nine Areas Combined	Eastern Cotton Area		
		Total	White	Negro
All ages	6.2	4.8	5.4	2.7
10 to 13 years	4.6	3.3	3.7	2.4
14 to 15 years	7.2	5.0	5.5	3.1
16 to 17 years	8.0	5.7	6.6	2.9
18 to 20 years	8.0	6.0	6.8	3.9
21 to 24 years	7.9	6.0	7.1	3.1
25 to 34 years	7.0	5.6	6.2	2.9
35 to 44 years	6.2	5.2	5.9	2.7
45 to 64 years	5.2	4.2	4.7	1.8

Source: Survey of Current Changes in the Rural Relief Population, Division of Social Research, W.P.A.

the 35 to 44 year group and two full grades beyond the 45 to 64 year group.

The relief and non-relief study referred to above yields information as to continuance in school (Table 50). Possibly most striking is the fact that only one-fourth (26 percent) of rural non-relief children 12 to 19 years of age had completed grade school. Among the Negroes this proportion dropped to less than 1 out of 10, and among the Negro relief cases to 1 out of 16. Comparatively few Negroes had completed high school.

Other studies show that the proportion of students in the rural Southeast who attend high school is relatively small. Of the total population 14 to 17 years of age in the United States, 31 percent were enrolled in high schools in rural communities in 1930, whereas of the southeastern cotton States, North Carolina alone rose higher than the national average with 33 percent. All other States of the Southeast fell below 25 percent,

---

[10] McCormick, T. C., *op. cit.*, Tables 25 and 26.

[11] McCormick, T. C., *idem.*

Alabama ranking at the bottom with less than 14 percent.[12] Comparison between urban and rural areas shows relatively more than twice as many students in high schools in urban areas, but the discrepancy tends to be much greater in the Southeast than elsewhere. For example, in Arkansas the proportion of children 14 to 17 years of age in rural communities enrolled in high schools in 1930 was less than 17 percent as compared with nearly 80 percent or almost five times as many in the urban schools of the State.[13] The development of consolidated schools in the Southeast is beginning to equalize urban-rural opportunities for whites, but few such schools are available for Negroes.

LENGTH OF TERM

Beyond the question of bare availability of schools, and directly related to attendance, is the question of the amount of education made available. The accompanying map and chart (Figure 30) show the deficiency of the cotton States in this respect. Unless schools are actually in session their presence or absence in the community does not greatly matter. Quoting Dr. W. H. Gaumnitz, Senior Specialist in Rural School Problems of the United States Office of Education: "School opportunities are obviously very different in a community where the school is open for 9 months from what they are where the school is open only 5 months. It is practically impossible to accomplish satisfactory results...in schools open less than half the year."[14] Yet all the southeastern cotton States were far below the average in length of rural school term in the United States (159.9 days) and three of them averaged less than 130 days in 1931-1932.[15] The discrepancy in length of urban and rural school terms is considerably greater in the Cotton South than elsewhere, the range being from a 59-day shorter rural term in Georgia to essentially equal terms in New Hampshire and New Jersey and a 6-day longer rural than urban term in Connecticut.

The situation with regard to Negroes was, of course, worse than that for whites. In the seven southeastern cotton States in 1931-1932 the average length of school term for Negroes was shorter than the term for whites, the difference ranging from 25 days in Georgia to 57 days in Louisiana.[16]

---

[12] *Economic Enrichment of the Small Secondary-School Curriculum*, Department of Rural Education, National Education Association, February 1934, p. 16.

[13] *Idem.*

[14] *Economic and Social Problems and Conditions of the Southern Appalachians*, U. S. Department of Agriculture, Miscellaneous Publication No. 205, January 1935, p. 99.

[15] *Biennial Survey of Education, 1930-1932, op. cit.*, p. 104.

[16] *Idem*, p. 95.

FIG. 30 - AVERAGE LENGTH OF SCHOOL TERM IN DAYS, 1927-1928

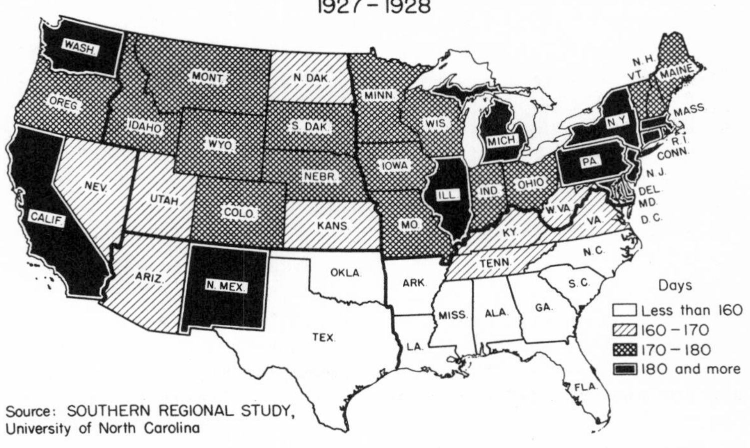

Source: SOUTHERN REGIONAL STUDY, University of North Carolina

AF-2037, W.P.A.

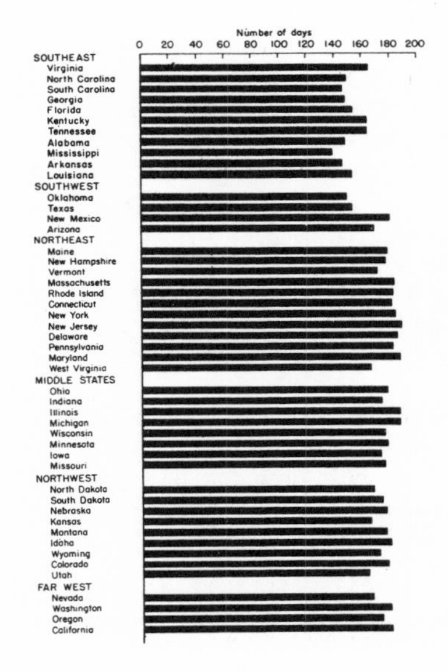

AVERAGE LENGTH OF SCHOOL TERM IN DAYS, 1927-1928

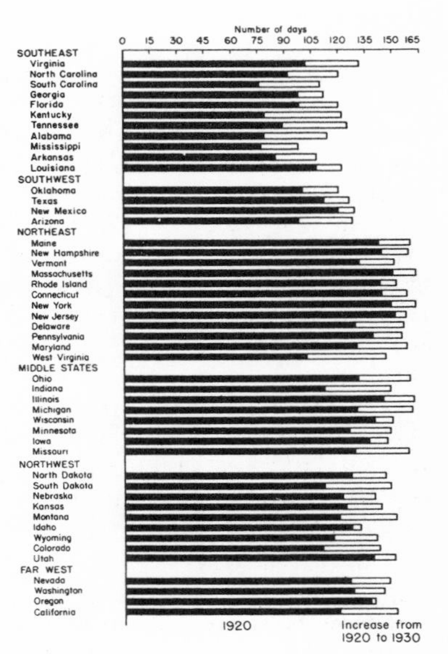

AVERAGE NUMBER OF DAYS ATTENDED BY EACH PUPIL ENROLLED IN 1920 and 1930

Source: SOUTHERN REGIONAL STUDY, University of North Carolina

AF-2039, W.P.A.

Recent computations made by the United States Office of Education[17] show that the average length of school term in 1929-1930 for rural one-teacher schools in the southern States was 166 days for white schools and 123 days for Negro schools. The variation among the States is considerable. Whereas in North Carolina and Georgia the Negro terms were only 5 and 6 days shorter, respectively, in South Carolina and Louisiana they were 66 and 77 days shorter. Moreover, South Carolina and Louisiana had the shortest terms for Negroes, with only 85 and 93 days respectively.

The average lengths of term of the various classes of white and Negro schools show that the greatest discrepancy comes in these one-teacher schools and also that they have the shortest terms. It is of obvious significance that the one-teacher schools are the ones which most directly concern the cotton families. The average lengths of term for the various classes of rural schools by race in 1929-1930 were as follows: one-teacher schools, white, 166 days, Negro, 123 days; two-teacher schools, white, 161 days, Negro, 128 days; schools of three or more teachers in open country, white, 165 days, Negro, 144 days; consolidated schools, white, 169 days, Negro, 156 days; schools of three or more teachers in villages and towns, white, 175 days, Negro, 158 days.[18] The depression increased the rural handicap, as many schools have been closed each year after 3 or 4 months operation, owing to lack of funds.

TEACHERS' TRAINING AND SALARIES

It is generally agreed that the factor of greatest importance in the school education of a child is his teacher. The salaries paid probably offer the best index as to the quality, training, and fitness of the persons employed for the task (Figure 31). Judged on this basis the rural United States fares badly compared with urban areas, as does the rural South compared with the rural United States as a whole. The average annual salary of rural teachers in the southeastern cotton States in 1931-1932 ranged from $485 in Arkansas to $702 in North Carolina while for urban teachers the range was from $967 in Arkansas to $1,287 in Louisiana, these averages being raised by the inclusion of supervisors and principals.[19]

The status of the rural teacher is well described in the following quotation from the National Inventory of Human Welfare: "During the school year 1933-34 one-half of all rural teachers

---

[17] Gaumnitz, Walter H., *Status of Teachers and Principals Employed in the Rural Schools of the United States*, Bulletin No. 3, 1932, p. 68.

[18] *Idem.*

[19] *Biennial Survey of Education, 1930-1932, op. cit.*, p. 107.

in the United States received less than $750 annual salary—less than the 'blanket code' minimum of the N.R.A. for unskilled labor! At least 40,000 of this low-salaried group received less than $500 a year. Many Negro teachers had an annual salary of as little as $100, and in agricultural sections experienced teachers were paid as low as $30 and $40 a month."[20] The five States with the lowest salaries paid to rural teachers were in the southeastern area.[21]

FIG. 31-AVERAGE ANNUAL SALARIES PAID PUBLIC SCHOOL TEACHERS, 1928

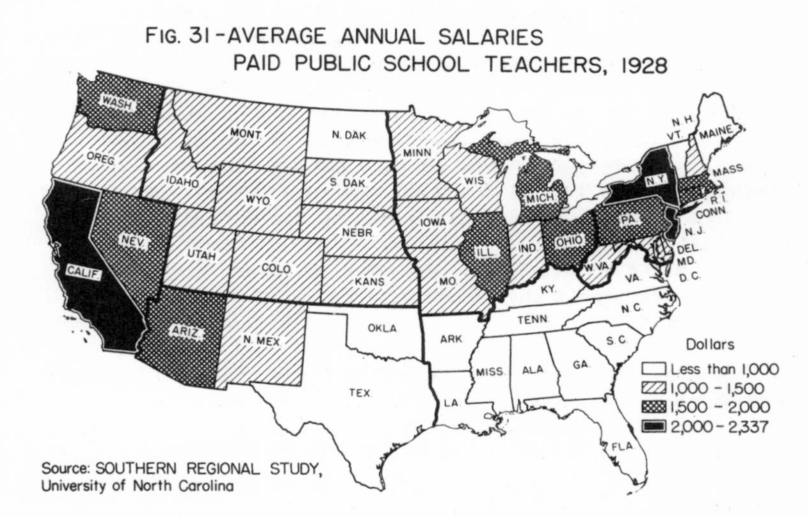

AF – 2041, W.P.A.

As noted above, the United States Office of Education recently published an extremely illuminating study[22] giving the salary, length of term, education, and experience of rural teachers by race and class of school. Comment here is restricted to but one item—salaries in one-teacher rural schools. Some very interesting facts are revealed by a comparison of the median salaries in the States. For example, the white teacher in the one-teacher schools of Georgia received an average (median) of $403 in 1929-1930, whereas the teacher in this same type of rural school in California received an average of $1,360, an amount more than three times as great. Of the southeastern States, 5

---

[20] *A National Inventory of Human Welfare*, No. 8, "Problems Which Confront the Public Schools." Information Service, Department of Research and Education, Federal Council of the Churches of Christ in America, February 8, 1936, Vol. XV, No. 6, p. 3.

[21] *Biennial Survey of Education, 1930-1932, op. cit.*

[22] Gaumnitz, Walter H., *op. cit.*

fell into the lowest 12 of the 48 States.[23] Apropos of the general range of these salaries Dr. Gaumnitz raised the following queries which are particularly suggestive when applied to the cotton States:

> How good a teacher and what quality of education can reasonably be expected for the salaries offered? Can a State with so little in the way of financial rewards hope to attract to its schools teachers who can perform adequately the very difficult, complex, and responsible task of assuming almost sole guidance of the educational development of rural children? Can high-grade young men and women under the salary conditions prevailing reasonably be expected to devote themselves seriously to the arduous and expensive task of obtaining a training commensurate with the task of teaching, and to a large degree, of supervising and administrating the work of these schools? Can their interest in rural teaching as a field of service be expected to be anything but transitory if the income offered is so unsatisfactory?[24]

The above comments refer only to white schools. The annual median salaries in the Negro one-teacher rural schools were also computed for the southern States. The States of Alabama, Georgia, Louisiana, Mississippi, and South Carolina all paid median salaries of less than $300. By way of comparing white and Negro education Dr. Gaumnitz made the following pertinent comments on the basis of these data:

> If the salary of the median white teacher employed in all the rural schools of the 17 [southern] States is computed it is found to be $788; for all the colored teachers employed in the rural schools of the same States, it is found to be $388, a differential of $400. By way of further comparison, the median salary of all classes of rural white teachers for the United States was found to be $945. These differences are significant and lead to some disturbing questions: Is it any wonder that Negro teachers as a group show particularly low training standards? What can be expected in the way of high-grade teaching performance when such meager bid is made for high-grade performers? Can we hope to

---

[23] Gaumnitz, Walter H., *op. cit.*, pp. 30-31.

[24] *Idem*, p. 32.

improve the public education provided for the Negro unless we are willing to put more into the making and retention of those charged with the important task of giving instruction? Can we logically expect the Negro race to fit into the American scheme of things socially, economically, and culturally if we continue to provide its constituents with an educational opportunity which at its mainspring, the teacher, is so seriously handicapped?[25]

Table 52—EXPENDITURES FOR TEACHERS' SALARIES IN PUBLIC SCHOOLS PER CHILD 6 - 14 YEARS OF AGE

State	Year	Per Capita		Percent of Increase	
		White	Negro	White	Negro
Oklahoma	1912-1913	$14.21	$ 9.96		
	1920-1921	41.94	24.85	195	149
Texas	1913-1914	10.08	5.74		
	1922-1923	32.45	14.35	222	150
Kentucky	1911-1912	8.13	8.53	-	-
Tennessee	1913-1914	8.27	4.83	-	-
North Carolina	1911-1912	5.27	2.02		
	1921-1922	26.74	10.03	407	397
Virginia	1911-1912	9.64	2.74		
	1921-1922	28.65	9.07	197	231
Arkansas	1912-1913	12.95	4.59		
	1921-1922	20.60	7.19	59	57
Louisiana	1911-1912	13.73	1.31		
	1921-1923	36.20	6.47	164	394
Florida	1910-1911	11.50	2.64		
	1921-1922	37.88	6.27	229	138
Georgia	1911-1912	9.58	1.76		
	1921-1922	23.68	5.54	147	215
Mississippi	1912-1913	10.60	2.26		
	1921-1922	28.41	4.42	168	96
Alabama	1911-1912	9.41	1.78		
	1921-1922	22.43	4.31	138	142
South Carolina	1911-1912	10.00	1.44		
	1921-1922	30.28	3.63	203	152

Source: Woofter, T. J., Jr., *The Basis of Racial Adjustment*, Boston, Ginn and Company, 1925, p. 178.

The limitations imposed upon the development of an adequate school system by low salaries for teachers can hardly be exaggerated. Not only do such salaries fail to attract adequate talent into rural teaching but they serve to drive many of the most efficient of the present teachers either into other lines of work, into urban schools, or into other States which pay better salaries. Thus, many of the better teachers are lost to the rural South because they are unable to maintain their self-respect and professional standing on the inadequate salaries offered them.

[25] Gaumnitz, Walter H., *op. cit.*, pp. 40-41.

The deficiency in teachers, however, is not entirely a matter of finance or unavailability of suitably trained persons. A recent Arkansas study compares the educational qualifications of 5,536 unemployed teachers who applied for relief teaching assignments between October and December of 1933 with the qualifications of 9,386 elementary teachers regularly employed in the State during 1932-1933.[26] The comparison indicates that those who were not employed and who made application for assignment on relief teaching had materially higher educational qualifications than those who were employed during the school year. More than 16 percent of the unemployed were college graduates as compared with less than 6 percent of the employed; more than 28 percent of the unemployed had had from 2 to 4 years of college training in comparison with 22 percent of the employed.

Table 53—PER CAPITA EXPENDITURE FOR TEACHERS' SALARIES IN COUNTIES BY PERCENT OF NEGROES IN THE TOTAL POPULATION

Percentage of Negroes in Total Population	Per Capita Expenditure for Teachers' Salaries	
	White	Negro
Counties under 10 percent Negro	$ 7.96	$7.23
Counties 10 to 25 percent Negro	9.55	5.55
Counties 25 to 50 percent Negro	11.11	3.19
Counties 50 to 75 percent Negro	12.53	1.77
Counties 75 percent and over	22.22	1.78

Source: U.S. Bureau of Education, Bulletin 38, 1916, p. 28.

Obviously, considerations other than educational qualifications often determine the choice of teachers. The school boards that select the teachers, and the members of the community that choose the school boards, are themselves victims of the educational inadequacies of the past.

Something of the educational situation existing when those who are now adults were growing up is shown in Table 52. These figures, while encouraging because they show substantial increase in per capita expenditure for teachers' salaries, are also discouraging when viewed as to actual amounts spent, particularly for Negro children. It is not surprising that Louisiana, for example, had a rural Negro illiteracy rate of 35 percent for the population 21 years of age and over in 1930 when back in 1911-1912, the date at which the average tenant farmer of today was of school age, the expenditure for teacher's salary per Negro child 6 to 14 years of age was but $1.31.

The following data (Table 53), selected from this same period, showing the per capita expenditure for teachers' salaries in counties grouped according to percentage of Negroes in the total population, show that the heavier the Negro population

[26] *Status of Common Schools and of Elementary Teachers in Arkansas, 1934*, Emergency Relief Administration of Arkansas (typed), p. 33.

the less spent for Negro teachers per capita, whereas the more is spent for white teachers per capita:[27]

> This means that in the country districts of these counties a few expensive schools are maintained for the scattered white pupils, while the congested Negroes can be herded into small one-teacher schools with wholly inadequate equipment. State school funds are distributed to these counties on the basis of their combined white and black school population or attendance.... The local school board then takes the State funds, adds a local tax and apportions it to white and colored schools as they please. Justice should demand that such funds be apportioned more closely in proportion to the population of the two races.[28]

Although substantial improvement has been made in this situation in the past 20 years the distribution of funds in many sections still leaves much to be desired.

FINANCIAL SUPPORT

The Southeast gives its schools less financial support than any other section of the United States both as regards capital outlay and allotments for current expense. With respect to the former item the rural schools of the country showed an average value of school property per pupil enrolled in 1931-1932 which was less than half that of the urban schools, the two figures being $143 and $353, respectively. The range for the rural schools was from $47 in Georgia to $460 in Nevada, all 7 southeastern cotton States falling within the lowest 12 States and none reaching as high as $100.[29]

One of the best measures of financial support is the per capita cost for current expenses and interest per pupil in average daily attendance. This gives the southeastern States a certain advantage in comparison with the other sections of the country in that the average daily attendance in this section is relatively low. But despite this all of the southeastern cotton States fall into the lowest 12 States of the Nation.[30] Georgia ranked lowest in 1931-1932 with an average expenditure of $25.93 in rural areas. The average for the rural United States was $64.39.

---

[27] *Status of Common Schools...in Arkansas, 1934, op. cit.*, p. 179.

[28] *Idem*, pp. 179-180.

[29] *Biennial Survey of Education, 1930-1932, op. cit.*, p. 109.

[30] *Idem*, pp. 109-110.

On the other hand, the Southeast is least able to pay for education on the basis of taxable property. A comparison of the tax value of the property with expenditures for schools shows that this section is spending as much for education per dollar of wealth as other regions. The southeastern States expended about the same proportion of their income for schools in 1928 as the national average, three of the cotton States being slightly below and three considerably above the average (Figure 32). However, this income was so small that in spite of their equality of effort the resultant appropriations for education were far below the norm for the country.

A POORER TYPE OF RURAL SCHOOL

PUBLIC ATTITUDE TOWARD EDUCATION

The various facts cited show clearly that public opinion in the Southeast is not squarely behind a progressive program of public education, especially Negro education. This has been illustrated again during the depression when it has been the schools that have suffered most from governmental economies.

The apathy of the public toward the establishment of a reasonable, not to say aggressive, program of public education, is to some extent a carry-over from the pre-Civil War period and to some extent springs from the same causes that produced this apathy under slave conditions. Literacy among the workers was not considered an asset to landowners in the plantation

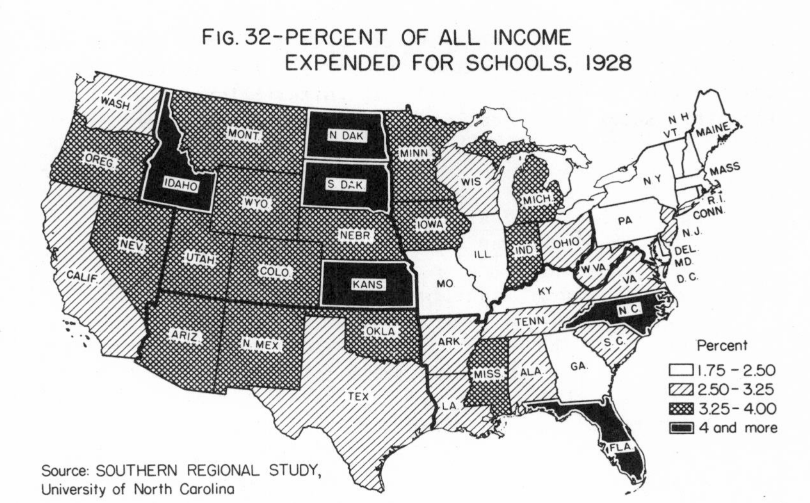

AF-2043, W.P.A.

DIFFERENTIALS IN REGIONAL EDUCATIONAL LOAD AND ECONOMIC CAPACITY

STATE	Percent Population Under 19 Years of Age 1930	Percent of Total Population Attending School 1930	Per Capita Income 1929	Per Capita Total Expenditures in Public Day Schools 1927-28	Percent School Expenditures are of all income 1928
Southeast					
Virginia	44.4	21.9	$ 74	$ 9.23	2.61
North Carolina	49.3	24.1	42	12.28	4.38
South Carolina	50.6	24.1	29	9.10	3.16
Georgia	46.3	21.7	56	6.11	1.75
Florida	39.2	19.7	117	20.76	5.76
Kentucky	43.9	21.5	74	8.58	2.29
Tennessee	43.8	21.9	69	8.72	2.57
Alabama	47.0	22.5	47	7.60	2.74
Mississippi	46.6	23.5	32	9.04	3.94
Arkansas	45.8	23.2	37	7.63	2.55
Louisiana	44.0	21.5	83	10.32	2.61
Southwest					
Oklahoma	44.2	23.7	92	12.24	3.27
Texas	42.6	21.1	94	11.32	2.57
New Mexico	46.8	22.7	74	12.36	3.40
Arizona	42.1	20.7	143	19.06	3.67
Northeast					
Maine	37.3	20.2	146	13.50	1.93
New Hampshire	35.2	19.4	160	14.42	2.14
Vermont	37.0	19.8	139	14.14	2.24
Massachusetts	35.1	20.3	326	19.55	1.85
Rhode Island	37.0	19.6	272	18.30	1.89
Connecticut	37.0	29.4	355	20.44	2.46
New York	33.6	19.0	506	23.86	2.11
New Jersey	36.1	19.9	308	25.82	3.20
Delaware	35.9	19.7	629	13.54	1.91
Pennsylvania	39.4	21.2	233	18.40	2.20
Maryland	37.2	19.1	264	13.20	1.97
West Virginia	46.1	23.1	80	14.91	3.21
Middle States					
Ohio	36.1	20.8	$192	$21.03	3.05
Indiana	36.5	20.3	114	22.04	3.93
Illinois	34.9	19.6	301	19.28	2.28
Michigan	37.7	20.7	219	22.57	3.92
Wisconsin	38.0	21.2	155	16.32	2.95
Minnesota	38.3	21.8	133	19.30	3.55
Iowa	37.2	21.1	90	20.18	3.82
Missouri	35.7	19.1	155	14.57	2.46
Northwest					
North Dakota	45.4	24.9	46	22.77	6.13
South Dakota	42.5	23.7	58	21.61	5.78
Nebraska	39.3	22.1	102	19.50	3.95
Kansas	38.1	21.9	97	22.82	4.24
Montana	39.0	22.8	133	24.24	3.96
Idaho	42.8	25.1	87	23.44	4.02
Wyoming	39.2	21.8	139	27.25	3.30
Colorado	38.0	21.6	155	24.02	3.29
Utah	46.1	27.1	119	21.17	3.91
Far West					
Nevada	31.8	18.1	241	25.24	3.33
Washington	33.7	21.2	186	57.91	2.80
Oregon	33.1	20.8	138	20.98	3.31
California	30.4	19.3	313	25.43	3.25
United States	38.8	20.9	?	17.77	2.74

economy preceding the Civil War. In fact, landowners regarded education of workers as a distinct disadvantage besides being an unnecessary expense. The conventional division of labor directed the planter to handle the accounts. Too much information on the part of the workers would be likely to bring them into conflict with the system, with resulting demoralization. Any attention which a worker gave to learning to read or write was regarded as evidence of his desire to escape from dependency. Accordingly schooling was not only discouraged, but was usually forbidden upon pain of severe punishment.

Somewhat the same situation exists today. The landowner, who keeps the accounts, believes that it is not to his advantage to encourage the education of his tenants or share-croppers. He discourages education especially for the Negroes, but also for the whites. Color does not particularly concern the plantation owner, who is interested in obtaining a certain type of worker who fits into the qualifications of the system, regardless of race.

The type of worker selected by landowners in general lacks initiative and aggressiveness, qualities which naturally affect his attitude toward education. Thus he contributes to his own educational retardation. The southern cotton tenant, particularly the cropper, commonly reflects the general apathy in the rural Southeast toward education. Only maximum interest in education could overcome the handicaps of excessive moving from place to place, short school terms, and economic disabilities. The tenant has no incentive to develop such interest. On the contrary he often keeps his children from school for what seem to be very trivial excuses. Of course, it is not always possible to determine the exact reasons why children are kept from school. Although no data are available on this point, it appears probable that the sickness rate is higher among share-croppers and tenants than in the general population. Among the poorer members of the group illnesses arising from improper diet and malnutrition may be the real reasons for the apparent apathy toward school attendance.

The labor system employed in cotton culture also militates against regular school attendance. Since returns per worker are small, the combined forces of entire families are required in the fields in peak seasons, which overlap the school term. To what extent this system interferes with education may be gauged by the fact that in 1930 in the 7 southeastern cotton States 31,764 children 10 to 15 years of age inclusive were occupied at agricultural labor as wage workers, while another 272,058 children of these ages were unpaid family workers.[31]

---

[31] *Fifteenth Census of the United States: 1930*, Occupations Vol. IV, Table 23.

School vacations during the heaviest of the picking season mitigate this influence to some extent. Regardless of the economic or other arguments used for or against child labor in the production of cotton, its retarding effects on the child's education cannot be questioned.

Another factor in retarding the educational development of the tenant child is the frequent moves made by the family. These moves usually cut into the school term since they commonly occur sometime between November and February after the cotton crop has been picked. The actual number of days lost from school on this account should not necessarily be great since the distance of such moves is usually short and the amount of goods to be moved offers no considerable obstacle. Changing schools in the middle of the year, however, breaks up the continuity of the school work and of the child's associations and not infrequently demoralizes the whole situation to the point of his being kept out of school entirely for the remainder of the school year.

Thus, it appears that the plantation system of cotton culture is not conducive to a high standard of public education. It has been noted (Figure 28) that where cotton is cultivated most extensively illiteracy rates are highest. A recent educational survey of Arkansas pointed out that of 12 counties in the State showing more than 10 percent illiteracy, 5 lead in acreages of cotton. The report further pointed out the relation among cotton, Negroes, and illiteracy, indicating that where cotton culture prospers the workers are likely to be illiterate Negroes. The counties involved have the highest ratio of colored population of all counties in the State.

> This makes it...scarcely possible to question... that the density of colored population and the dominance of cotton...are economically allied, and that the condition of the economic status attendant on such alliance is responsible for the illiteracy... Any plans looking toward an immediate amelioration of the condition of illiteracy in these areas cannot be worthy of mature consideration unless they take into consideration...an improvement of the economic alliance above referred to.[32]

The old plantation system thrived on masses of ignorant workers. The new order in agriculture, necessary to rehabilitate the South, gives rise to entirely different problems, and these modern agricultural problems cannot be successfully met with

---

[32] *Status of Common Schools and of Elementary Teachers in Arkansas, 1934*, Emergency Relief Administration of Arkansas (typed), p. 4.

ignorance. Yet the schools, the chief agencies by which ignorance may be dispelled, are allowed to languish for want of interest and adequate financial support. The ignorance and shortsightedness of the cotton share-cropper of this generation is but a reflection of his training in the past. In like manner succeeding generations of adults will be handicapped by the inadequacies of their childhood training unless a new order of rural education can be effected.

Chapter X

RELIEF AND REHABILITATION

The very nature of the share-cropping system presents a situation in the rural South unlike that faced elsewhere in the administration of relief. A review of the preceding chapters reveals a number of reasons for the relatively high rural relief rate in the southeastern States in the early months of the relief program, and the relatively low rate in the later months of the program. The prevailing one-crop system of farming with consequent soil depletion, the cropper system with the concomitant low standard of living and high illiteracy rate, the advent of the weevil, and the expansion of southwestern cotton production set the stage for the prolonged agricultural depression with frequent periods of low cotton prices. Disorganization of agriculture resulted in curtailment of operations by some planters and absolute cessation of planting by others between 1930 and 1932. Furthermore, tobacco acreage was decreased by about one-third during this period. The result was a displacement of large numbers of cropper families during these years.[1] Rural youth, no longer able to get employment in urban centers and unable to gain entrance to the agricultural economy of the South, augmented unemployment rolls. Land foreclosures increased enormously (chapter II). Furthermore, an important non-agricultural cause of the high relief rates in some farming areas of the Southeast was the decadence of lumbering and naval stores industries which had formerly provided employment for farmers in off-seasons.

EXTENT AND TREND OF RELIEF

These conditions multiplied distress in the seven southeastern cotton States to the point that half a million cases were on the relief rolls early in the program (October 1933) (Appendix Table 56). This constituted slightly more than one-eighth of the number of families in the area as reported in the 1930 Census. Of the total half million cases in the cotton States

---

[1] Blackwell, Gordon W., "The Displaced Tenant Farm Family in North Carolina", *Social Forces*, October 1934, Vol. 13, No. 1. Sample counties were Greene, Nash, and Wilson. Also, Beck, P. G. and Forster, M. C., *Six Rural Problem Areas, Relief-Resources-Rehabilitation*, F.E.R.A. Research Monograph I, 1935, pp. 63-64.

in October 1933, almost 300,000 were in cotton counties.[2] Over 200,000 of these were rural cases, and many of the urban clients had followed occupations whose prosperity was dependent upon cotton. The relative incidence of relief in cotton and non-cotton counties, in urban and rural areas, and in white and Negro families is shown in Figure 33.

The number of relief cases continued to increase through the winter of 1933-1934 until, in January 1934, they amounted to more than one-sixth of the 1930 families, in spite of the transfer of thousands of families from relief to the Civil Works Administration (Appendix Table 57 and Figure 34). At this time the relief rate in the Cotton Plantation Belt was higher than in any other major agricultural area of the United States except the Spring Wheat Area.[3]

In the spring of 1934 the rural rehabilitation program was inaugurated with the objective of removing farmers from the dole and aiding them to attain self-support. Subsequent appraisal of the relief situation must, therefore, be based on a combination of relief and rehabilitation trends.[4]

From a peak of 293,000 cases in November 1933, the number of rural relief cases in cotton counties had dropped to 251,000 in May 1934, 13,000 cases having been transferred to rural rehabilitation. For a year the number of relief and rehabilitation cases combined remained about the same, declining only 5 percent by May 1935. During this 12 months, however, 42,000 cases were transferred to rehabilitation (Appendix Table 58 and Figure 35). Between May 1935 and November 1935 the combined total decreased by 117,000, due partly to the Works Program.

While the decrease in relief in cotton counties between May 1934 and May 1935 was largely balanced by the increase in families on rural rehabilitation rolls, the shift does not hold true for each individual State. In the cotton counties of Arkansas and Georgia, relief and rehabilitation cases both increased. In Alabama, Mississippi, and South Carolina the relief load decreased far more than the rehabilitation load increased. In North Carolina slightly more cases were taken on rehabilitation

---

[2] In order to focus the discussion more closely on the plantation, those counties with 40 percent or more of their gross farm income from cotton were tabulated separately from the tobacco, general farming, self-sufficing, and other crop specialty counties.

[3] See Mangus, A. R., *Changing Aspects of Rural Relief*, and Asch, Berta, *Farm Families on Relief and Rehabilitation*, forthcoming monographs of the Division of Social Research, W.P.A.

[4] Families under the rural rehabilitation program were for the most part former relief clients still dependent upon government assistance. They are, therefore, included with relief families in a combined intensity rate. While this provides an accurate picture of the trend, it causes some duplication in the total number of cases as some rehabilitation clients received supplementary aid from work relief and consequently are counted in both programs.

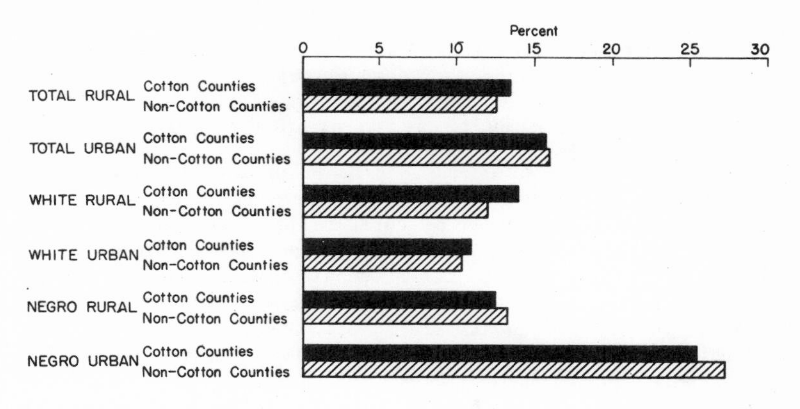

FIG. 33–RELIEF RATES IN THE SEVEN SOUTHEASTERN COTTON STATES BY RESIDENCE, COLOR, AND LOCATION IN COTTON OR NON-COTTON COUNTIES, October 1933

Source: Unemployment Relief Census, October 1933, Report Number Two, Table 9, pp. 106–211

AF–1451, W.P.A.

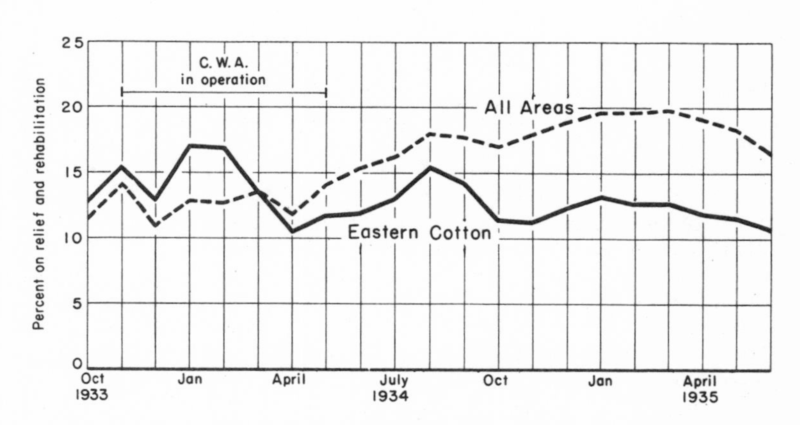

FIG. 34–TREND OF COMBINED RELIEF AND REHABILITATION IN RURAL COUNTIES BY AREAS* October 1933 through June 1935

* Survey of current changes in the rural relief population

Source: Division of Research, Statistics and Finance, F.E.R.A.

AF–1419, W.P.A.

than were dropped from relief while in Louisiana both relief and rehabilitation cases decreased slightly.

FACTORS INFLUENCING RELIEF TRENDS

The ebb and rise of economic prosperity are interwoven with administrative policy and racial differences in effecting the fluctuations in the number of rural relief cases in the Southeast. In the first years of the depression many landlords could no longer furnish food to their tenants. Some sought to shift this burden to the Emergency Relief Administration and were successful in certain localities. Soon, however, administrative policies were formulated to minimize this practice and relief loads were reduced accordingly. For example, the reduction of the relief load in cotton counties in Louisiana by more than one-half between November 1933 and May 1934 was largely due to an administrative order from the State office removing from relief rolls all tenants on plantations.[5]

The rise in the price of cotton following the inauguration of the Agricultural Adjustment Administration crop control program made it possible for a considerable number of farm operators to leave the relief rolls. Although the small amount of benefit payments received by tenants can hardly be considered a factor in the reduction of the rural relief rate—since more often than not benefit payments to croppers were applied by landlords on back debts accumulated during the depression—the rise in cotton prices enabled many landlords again to extend credit for subsistence to tenants and thus indirectly resulted in removing families from relief rolls.

Governmental agencies to aid the farmer reached the farm operator on relief much less frequently than they did his neighbor who was fortunate enough to stay off relief. A study of rural relief and non-relief households in October 1933 revealed that 62 percent of non-relief farm operators studied in the Old South Cotton area received A.A.A. benefit payments, while this was true of only 31 percent of the farm operators on relief in the area. Similarly 14 percent of the non-relief farm operators received assistance from the Farm Credit Administration, while loans were made by this agency to only 9 percent of the farm operators on relief.[6]

Partial, if indirect, credit due the A.A.A. for reduction of the relief load is indicated by data presented in Table 54 and Figure 36. In general, counties which had received the largest

---

[5]The relief rate in the 32 cotton counties in Louisiana decreased from 12.7 in November 1933 to 6.0 in May 1934. Source: Division of Research, Statistics, and Finance, F.E.R.A.

[6]McCormick, T. C., *Comparative Study of Rural Relief and Non-Relief Households*, W.P.A. Research Monograph II, 1935, Table 11, p. 83.

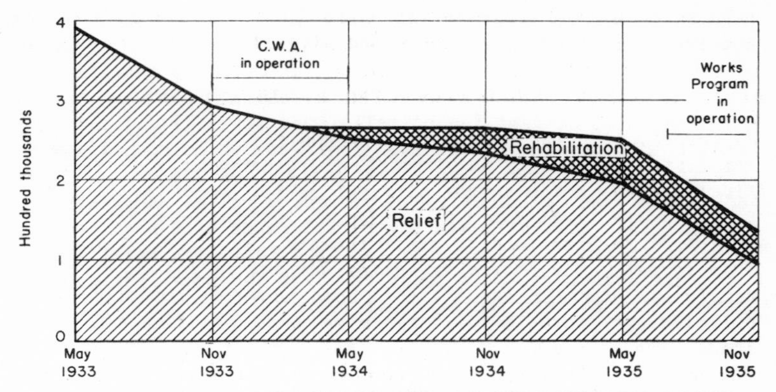

FIG. 35 – NUMBER OF RELIEF AND REHABILITATION CASES IN COTTON COUNTIES IN SEVEN SOUTHEASTERN COTTON STATES BY SIX-MONTH INTERVALS

May 1933 through November 1935

Source: Division of Research, Statistics, and Finance, F.E.R.A. and Resettlement Administration

AF – 1413, W. P. A.

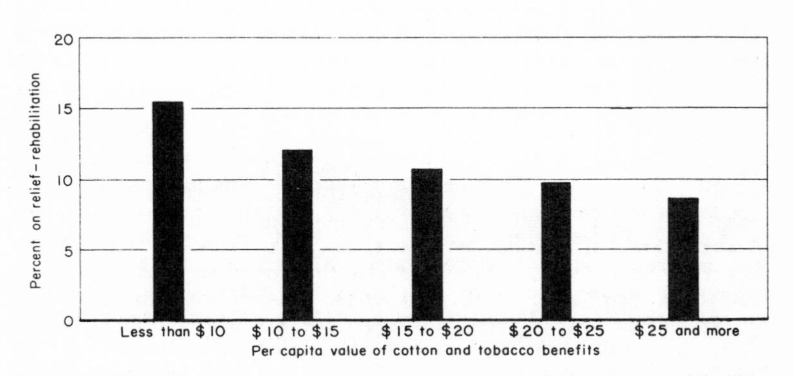

FIG. 36 – MEDIAN RELIEF-REHABILITATION RATES, MAY 1935, IN COTTON AND TOBACCO COUNTIES OF SEVEN SOUTHEASTERN COTTON STATES, BY PER CAPITA VALUE OF COTTON BENEFIT PAYMENTS

September 1933 through March 1936

Source: Division of Research, Statistics, and Finance, F.E.R.A. and Agricultural Adjustment Administration

AF – 1415, W. P. A.

per capita amount of cotton and tobacco benefits since the inauguration of A.A.A. had the lowest relief and rehabilitation rates in May 1935. Those counties which received benefit payments amounting to an average of less than $10 per capita had a median ratio of families on relief and rehabilitation in May 1935 of 15.5 percent of the 1930 families, while counties with benefit payments amounting to an average of $25 or more per capita had a median of only 8.6 percent on relief and rehabilitation. This association of low relief rates with high benefit payments is also influenced by the fact that high benefit payments are concentrated in counties where the prevalent cropper system tends to keep families off relief.

Table 54—COTTON AND TOBACCO COUNTIES[a] IN SEVEN SOUTHEASTERN COTTON STATES[b] BY PERCENT OF FAMILIES ON RELIEF AND REHABILITATION[c] AND BY PER CAPITA CROP BENEFIT PAYMENTS[d]

Percent of 1930 Families on Relief and Rehabilitation May 1935	Per Capita Cotton and Tobacco Benefits (1933 through March 1936)						
	Total Counties	Less than $10	$10–$15	$15–$20	$20–$25	$25 and over	Median Value
Total	201	28	50	65	31	27	$16.73
Less than 7.5	44	2	7	15	10	10	19.33
7.5 to 10	46	5	13	14	6	8	16.79
10 to 15	50	6	12	24	4	4	16.46
15 to 20	38	10	11	6	8	3	14.09
20 and over	23	5	7	6	3	2	14.64
Median percent	11.1	15.5	12.1	10.7	9.8	8.6	

[a] A cotton county is one in which 40 percent or more of the gross farm income in 1929 came from cotton farms. A tobacco county is one in which 10 percent or more of the cultivated acreage in 1929 was planted in tobacco. Greene County, Alabama, is excluded because no Federal relief or rehabilitation program was operative during May 1935 in the county.

[b] Alabama, Arkansas, Georgia, Louisiana, Mississippi, North Carolina, and South Carolina.

[c] Information from Division of Research, Statistics, and Finance, F.E.R.A.

[d] Information from A.A.A. mimeographed releases by counties and *Fifteenth Census of the United States: 1930*.

Administrative factors which were influential in determining relief loads included attitudes of landlords and tenants toward relief, consideration of these attitudes by administrators, and the funds and personnel available for meeting the needs.

A study in December 1933 of 1,022 farm families on relief in Alabama revealed that approximately 30 percent of the tenants receiving assistance were helped by their landlords to get on relief rolls.[7] On the other hand, many planters viewed relief with suspicion because of its alleged "demoralizing" effects. Many landlords doubtless feared that relief and rehabilitation would possibly result in stirring up a hitherto docile labor supply. Acquiescing to landlords' requests, relief offices in many rural cotton counties in the deep South were closed during the 2 months of cotton picking. Thus, the labor supply for cotton picking was augmented and standards of relief expenditures

---

[7] Hoffsommer, Harold, *Landlord-Tenant Relations and Relief in Alabama*, Research Bulletin Series II, No. 9, Division of Research, Statistics, and Finance, F.E.R.A., November 14, 1935.

prevented from disturbing the local rate of wages for cotton pickers.

The adoption of more stringent rules for receipt of relief partially explains the decreasing relief rate in the rural South.[8] Although new families were continually being forced to ask for assistance, a number of clients were being dropped as they became self-supporting. In most of the southeastern States there was little or no precedent upon which to rely in determining relief needs. Prior to the fall of 1932, comparatively little had been done on a State-wide basis in these States in the field of public welfare and social work. Obtaining trained personnel for positions in rural county relief offices was extremely difficult. Experience in relief administration during 1933 and the development of more capable case work personnel enabled some of these States to weed out their relief loads considerably.

Up to 1935 only two of the seven southeastern cotton States had State welfare departments with mandatory provisions for county welfare work, and one of these agencies was limited to child welfare.[9] When the depression set in, county administrative units in the South were little disposed and frequently unable to increase their appropriations for assistance to the needy. The result was that early in the relief program a greater proportion of the emergency relief burden, including a large percent of the so-called unemployables, had to be borne by the Federal Emergency Relief Administration in the cotton States than in most other States.

In October 1934 a study of 11 agricultural areas revealed that the Eastern Cotton Belt, with 14 percent of its rural and town relief cases having no worker, had a larger proportion of unemployables on relief than any other area, with 3 exceptions.[10] This high rate of unemployability can be explained partially by the relatively high proportion of unemployables among Negro relief cases. During the period July 1934 through June 1935, however, F.E.R.A. made a determined effort to shift unemployables to the care of administrative county units in the South. An important factor in the relatively low combined rural relief and rehabilitation rate in the Eastern Cotton Belt was this removal

---

[8]Beck, P. G. and Forster, M. C., *op. cit.*, p. 29.

[9]The two State agencies were the North Carolina State Board of Charities and Public Welfare, created in 1917, and the Alabama Child Welfare Department, created in 1919. For legislative trends in the seven States, see Lowe, R. C., *Legislative Trends in Public Relief and Assistance, December 31, 1929 to July 1, 1936*, Division of Social Research, W.P.A., Series III, No. 2, (to be published); also *Digests of Public Welfare Provisions under the Laws of Alabama, Arkansas, Georgia, Louisiana, Mississippi, North Carolina, and South Carolina*, Division of Social Research, W.P.A.

[10]Standing, T. G. and Mangus, A. R., *Workers and Dependent Age Groups in Rural and Town Relief Cases in October 1934*, Research Bulletin F-6, Division of Research, Statistics, and Finance, F.E.R.A., April 8, 1935.

of cases for administrative reasons,[11] although, as pointed out above, little local relief was available.[12] During the period October 1934 to February 1935 the number of rural relief cases with no employable member decreased in this area by almost three-fifths. At the same time five of the eight other agricultural areas studied showed much smaller decreases, while three actually showed increases in the number of unemployables on relief rolls.

The racial factor is more important in the Eastern Cotton Belt than in other rural areas. Since the effects of racial attitudes have always been felt by Negroes in the South in all phases of life, it is only to be expected that these attitudes would influence the administration of relief. The prevailing policy in some localities has been that a Negro must be in much more desperate straits than a white person before he can qualify for relief assistance. Owing to the nature of Negro-white relationships, a Negro would be much more hesitant than a white man in pushing his request for assistance if denied relief. In October 1933, three of the large cotton producing States in the Southeast—Mississippi, Arkansas, and Louisiana—had a Negro rural relief rate approximately one-half that for whites.[13] For this same month the relief rate for Negroes in rural areas of cotton counties in all seven States of the Southeast was slightly lower than that for whites (Figure 33), whereas in other areas Negroes were over-represented. In February 1935, the ratio of Negro families to all families in the Eastern Cotton Area was about 6 percent less in the relief population than in the general population of 1930.[14]

All of the under-representation of Negroes on rural relief rolls cannot be explained by discrimination. There are other factors which account in part for the lower Negro relief rate. Many cotton planters prefer Negro to white tenants.[15] Hence, Negroes, more than whites, may have retained the economic protection of the paternalistic tenure system through the depression.

---

[11] For example, see McGill, K. H., Hayes, Grant and Farnham, Rebecca, *Survey of Cases Removed from Relief Rolls in Seventeen Rural Counties in Georgia for Administrative Reasons in May and June 1935*, Research Bulletin Series II, No. 8, Division of Relief, Statistics, and Finance, F.E.R.A., November 4, 1935.

[12] Mangus, A. R., *The Trend of Rural Relief, October 1933—October 1934*, Research Bulletin F-3, Division of Research, Statistics, and Finance, F.E.R.A., March 22, 1935, p. 1; also Hulett, J. E., Jr., *Some Types of Unemployability in Rural Relief Cases, February 1935*, Research Bulletin H-2, Division of Research, Statistics, and Finance, F.E.R.A., October 4, 1935.

[13] *Unemployment Relief Census, October 1933*, Division of Research, Statistics, and Finance, F.E.R.A., Report Number Two, Table B, p. 14.

[14] Mangus, A. R., *The Rural Negro on Relief, February 1935*, Research Bulletin H-3, Division of Research, Statistics, and Finance, F.E.R.A., October 17, 1935, p. 1.

[15] Hoffsommer, Harold, *op. cit.*, p. 8.

Also the prevailing standard of living of Negroes in rural areas of the South has been lower than that for whites. To sink below this level and thus become eligible for relief was difficult. Furthermore, it may be pointed out that Negro tenants in the South tend to be concentrated in the rich soil belts in which the commercial plantation system holds sway.[16] Of all tenants in the seven southeastern cotton States, 53 percent were Negro in 1930, while of plantation tenants included in this study 84 percent were Negro. As will be shown later in this chapter, the relief and rehabilitation rate among plantation families has been considerably lower than that among farm families in general in the Eastern Cotton Area, owing perhaps to the fact that displacement of tenants has been less frequent on plantations.[17] Evidently those tenants located on small, individual farms on the fringes of the Cotton Belt have been more likely to apply for relief than have families on plantations. The concentration of Negroes on plantations, with whites more generally on small farming units, is, therefore, an important explanation of the under-representation of Negroes on rural relief rolls in this area.

NATURE AND EXTENT OF THE DISPLACED TENANT PROBLEM

Much of the public discussion relating to the effect of governmental programs on southern agriculture has centered around the displacement of tenants by acreage reduction. This topic, therefore, deserves more detailed analysis than the other factors affecting the relief picture.

A displaced tenant as referred to in this section is a former cropper or other share tenant who no longer has a cropping agreement with a landlord, *i.e.*, one who no longer receives advances for food and fertilizer for the production of a money crop, or the use of a work animal except for casual plowing. He may not be physically displaced in the sense of having been evicted from the plantation. In other studies it has been found that many tenants who could not be furnished by the landowner were still allowed to live, rent free, on the premises and were also allowed to use patches of land for cultivation of subsistence crops, but few of these were living on plantations enumerated in this study,[18] displacement probably being concentrated

---

[16] As noted in the Introduction, the average yield of cotton per acre is much higher on plantations than on all farms in the Southeast, being 253 and 216 pounds, respectively, in 1934. On plantations operated by wage hands, croppers, or other share tenants (excluding renters), the average yield was 257 pounds (Table 35, chapter VI).

[17] See next section, Nature and Extent of the Displaced Tenant Problem.

[18] Only 49 displaced tenants were enumerated in comparison with a total of 3,602 cropper and other share tenant families enumerated.

on smaller tracts and plantations where operations were abandoned. As a consideration of such an arrangement it was often understood that the displaced tenant would work for the landlord when his services were needed for odd jobs, such as fencing, ditching, cotton chopping, or picking. Thus, the displaced tenant was in reality reduced to the status of a casual laborer with rent free perquisites.

The measurement of the extent and character of this dislocation of the tenant population is difficult, since such operators, if they cultivate as much as three acres of land, even in the most casual manner, are classed by the Census as tenant farmers. It must also be kept in mind that the net number of tenants may remain stationary or actually show a slight increase, yet there may be a considerable displacement of older or less desirable farmers by younger and more vigorous men.

In addition, the Census of 1935 enumerated as tenants many families whose heads would fit the definition of displaced tenants given above.[19] If they cultivated as much as three acres of land, they fulfilled the Census definition of a tenant, but in many cases this three acres or slightly more was in sketchily cultivated subsistence crops and no regular money crop agreement was in force.

Although about the same number of tenants was recorded in the 1935 Census of Agriculture as in that of 1930 in the seven southeastern cotton States, the figures indicate a rapid rate of turnover, a shift from Negro to white tenancy and much displacement in restricted areas, offset by increases in other areas. As pointed out above, this displacement occurred mainly before the inauguration of the A.A.A. program.

The shifts in tenancy between 1930 and 1935 are apparent when tenants are segregated by color, by type of tenure, and by cotton and non-cotton counties.

The color shift from 1930 to 1935 was a continuation of the marked growth of white tenancy from 1910 to 1930. While the number of white farm operators was increasing from 1930 to 1935 in the 7 southeastern cotton States by 108,000 (10.9 percent), there was a decrease of 50,000 (-7.5 percent) among Negro farm operators (Appendix Table 59). The number of Negro full owners increased slightly, this increase being partially offset by a decrease in part owners. The tremendous decrease among Negro farm operators occurred in the "other tenant" status in which there were 40,000 fewer tenants in 1935 than in 1930. On the other hand, the decrease in the number of Negro croppers was relatively smaller than that of white croppers.[20]

---

[19] Census schedules were inspected to verify this point.

[20] For analysis of Negro-white differences in farm tenure shifts, 1930-1935, in North Carolina, see Hobbs, S. H., Jr., the *University of North Carolina Newsletter*, December 5, 1935.

Shifts in the status of the tenant group have been influenced decidedly by the cotton and tobacco economy.[21] Croppers decreased relatively more in non-cotton and non-tobacco counties than in cotton and tobacco counties. In contrast other tenants increased markedly in non-cotton and non-tobacco counties (Appendix Table 60). The result is that the number of all tenants in cotton and tobacco counties decreased slightly while in non-cotton and non-tobacco counties the number of all tenants increased considerably.

A larger proportion of cotton and tobacco counties show a marked decrease (more than 5 percent) between 1930 and 1935 in the number of both croppers and other tenants than is the case with non-cotton and non-tobacco counties (Appendix Table 61). Croppers decreased 5 percent or more in 53 percent of the cotton and tobacco counties and in 44 percent of the non-cotton and non-tobacco counties. The net decrease in the number of croppers in cotton and tobacco counties ranged as high as 52 percent in one county. The difference between cotton-tobacco and non-cotton-tobacco counties is even more striking for other tenants, the percentage of counties showing noticeable decreases being 27 and 9, respectively. It should be noted that the great majority of non-cotton and non-tobacco counties showing decreases in the total number of tenants are located in Georgia where the boll weevil and soil erosion have resulted in a decided contraction in cotton production. Most of these were cotton counties until the last two decades.

The nature of the displaced tenant problem in commercial farming areas, where there had been a noticeable decrease in the number of croppers, is indicated in a study carried on early in 1934 of 825 displaced tenant farm families in eastern North Carolina.[22] Very few of the families included in this study had been displaced before 1929. Three-fifths had lost farm operator status during the years 1929 to 1932. In 1933 the number of displacements decreased sharply, only to rise again slightly in 1934. The years 1931 and 1932 constituted the peak period. Negroes seem to have been displaced earlier than whites. A later North Carolina study of 142 displaced tenants[23] revealed

---

[21] In discussing shifts in tenancy in the Southeast, tobacco as well as cotton must be reckoned with. A tobacco county is defined as one in which 10 percent or more of the 1930 crop acreage was planted in tobacco. There are 31 tobacco counties in North Carolina (5 of which are also cotton counties as defined above), 5 in South Carolina (1 of which is also a cotton county), and 3 in Georgia. In the 7 States there are 367 cotton and tobacco counties and 226 non-cotton and non-tobacco counties.

[22] Blackwell, Gordon W., *op.cit.*

[23] Unpublished data from Study of Rural Relief Families in North Carolina by Gordon W. Blackwell for the North Carolina Emergency Relief Administration, 1934. Sample counties included in this tabulation were Bertie, Greene, Stokes, and Washington.

the same trend in displacement with a larger proportion being displaced in 1934.

An analysis of the reasons why tenants were displaced, as reported by both tenant and landlord, indicates that the most important cause was the inability of landlords at the depth of the depression to continue to finance all their tenants. They were forced to cut down expenditures which meant that they could not advance subsistence and production credit to as many families as formerly.

Farm mortgage foreclosures also resulted in many cropper families becoming unemployed. As noted in chapter II (Appendix Table 8), approximately 10 percent of the land in the southeastern cotton States has been taken over in recent years by land banks, depository banks, insurance companies, and other corporations. County and State governments have taken over farms for delinquent taxes, although it has usually been the policy of governmental units to do this only as a last resort. Sometimes foreclosed land has been rented to tenants who could furnish themselves,[24] the resident cropper families being evicted. In some instances the land has remained idle and tenants allowed to live in the houses and subsist on their own resources.

Another cause for displacement of tenants has been a shifting from farming with croppers to farming with wage hands by some landlords. This practice seems to be localized in certain areas, principally in the Mississippi Delta country. One county farm demonstration agent, himself long a cotton farmer of the Upper Delta, summed up the situation in 1935 as follows: "Sure I'm going to shift to farming entirely by day labor next year. It's the only way a landlord can make money now."

It may be asked why, with the rise in the price of cotton, displaced croppers have not been re-employed. It appears that the A.A.A. to some extent "froze" the number of croppers employed in cotton culture in the Southeast at something like the 1932 figure. Cotton acreage allowed under the A.A.A. reduction program was based upon acreage planted during the depression years. It was during these years that croppers were being displaced as landlords were unable to finance them and as cotton production was being curtailed in the Southeast and expanded in the Southwest. Thus, the acreage reduction program of the A.A.A. in effect barred the return of thousands of families into the money crop tenant class in the Eastern Cotton Area. As pointed out by the Brookings report on *Cotton and the A.A.A.*,[25] the acreage reduction program has resulted in relatively little net displacement of croppers since 1932 except in a few areas.

---

[24] Certain land banks and insurance companies adhere strictly to this rule.
[25] Richards, Henry I., The Brookings Institution, Washington, D. C., 1936, pp. 150-162.

However, without arguing the merits or flaws in the program, the fact must be faced that thousands of families formerly employed as cotton croppers cannot get a cotton crop to tend so long as production is restricted. Youth recently coming of age have had little chance to enter cotton production except as casual laborers and have often been partially dependent upon public assistance. If a displaced tenant or a youth is successful in getting a crop, some other cropper usually finds himself adrift.

Displacement of cotton tenants seems to have taken place largely from small farming units rather than from plantations.

Table 55—PERCENT CHANGE IN NUMBER OF PLANTATION FAMILIES, 1930-1935, BY COLOR AND BY TENURE STATUS, BY AREAS
(Cotton Plantation Enumeration)

Area	Total Plantations Reporting[a]	Percentage Increase or Decrease 1930-1935							
		Total	Color		Tenure				
			White	Negro	Wage Hands	Croppers	Share Tenants	Renters	Displaced Tenants
Total	526	7.9	36.5	5.2	19.5	1.4	6.0	18.7	157.8
Atlantic Coast Plain	40	20.8	26.5	19.8	29.7	- 3.7	d	d	d
Upper Piedmont	37	14.4	22.3	2.4	38.0	-12.1	50.0	d	d
Black Belt (A)[b]	99	12.1	54.1	6.8	6.7	7.0	20.5	29.6	d
Black Belt (B)[c]	9C	24.8	30.2	24.2	22.8	31.7	- 3.4	18.9	d
Upper Delta	82	5.4	34.0	4.7	- 4.0	7.2	- 0.3	d	d
Lower Delta	46	3.3	d	2.6	d	- 2.3	- 7.5	4.5	d
Muscle Shoals	19	27.9	42.6	12.0	d	d	12.8	21.9	–
Interior Plain	29	7.4	70.6	1.4	-12.8	6.9	9.9	–	d
Mississippi Bluffs	35	16.1	80.8	12.6	23.7	12.6	20.5	18.9	d
Red River	25	-4.6	16.4	-7.3	40.2	-26.5	5.2	–	–
Arkansas River	24	-3.6	38.9	-5.8	55.9	- 9.8	-10.6	d	–

[a] Data not available for 120 plantations; 1930 base is 7,108 families of whom 616 were white and 6,492 were Negro and of whom 1,107 were wage hands, 4,290 croppers, 973 share tenants, 674 renters and 64 displaced tenants.

[b] Cropper and other share tenant majority.

[c] Renter majority.

[d] 1930 base less than 25 families.

The 40 counties included in this study of plantations show a decrease from 1930 to 1935 of approximately 9,000 tenants, 5,000 of whom were croppers. This means that there were 7 percent fewer tenants in these counties in 1935 than in 1930. However, the number of families on plantations included in the study *increased* from 1930 to 1935 by approximately 8 percent (Table 55). In only two areas, Red River and Arkansas River, was there a slight decrease. For the most part the increase was distributed fairly evenly over the 5-year period. Increase in the number of Negro families was slight, while the number of white families increased by more than one-third. The greatest increases in the number of plantation families occurred in the wage hand and renter groups. Croppers and share tenants increased only slightly. The relatively small group of displaced tenants increased from 64 to 165 in number.

Changes in crop acreage may partially explain increases or decreases in the number of plantation families employed

(Table 56). Considering all areas together there was practically no change in the amount of crop acreage on plantations. Only in the Red River and Arkansas River areas were there significant decreases, an average of 41 and 65 acres less per plantation in the respective areas. Also, these were the only areas showing a decrease in the number of families.

Cotton plantations were probably employing more families in 1935 than in 1930, with the exception of a few specific localities. The only indication available as to whether as many families were being employed on cotton plantations in 1935 as during the period 1920 to 1930 is the number of vacant houses[26] and the amount of idle crop land (Table 57). Of the 646 plantations, 180 or more than one-fourth, reported an average of

Table 56—CHANGES IN CROP ACREAGE, 1930-1935, ON 513 PLANTATIONS, BY AREAS
(Cotton Plantation Enumeration)

Area	Total Plantations Reporting[a]	Number Acres Increase or Decrease 1930-1935	Percent Increase or Decrease	Number Acres Increase or Decrease per Plantation
Total	513	733	0.4	1.4
Atlantic Coast Plain	37	478	4.1	12.9
Upper Piedmont	36	- 195	-2.5	- 5.4
Black Belt (A)[b]	99	362	1.3	3.7
Black Belt (B)[c]	91	2,001	8.7	22.0
Upper Delta	84	457	0.9	5.4
Lower Delta	44	- 243	-2.6	- 5.5
Muscle Shoals	19	53	1.2	2.8
Interior Plain	28	- 289	-1.8	-10.3
Mississippi Bluffs	34	216	1.7	6.4
Red River	23	- 934	-7.3	-40.6
Arkansas River	18	-1,173	-6.8	-65.2

[a]Data not available for 133 plantations.
[b]Cropper and other share tenant majority.
[c]Renter majority.

2.7 vacant houses. They were reported most frequently in the Arkansas River and renter-majority Black Belt areas. The Upper Delta and Arkansas River areas had the largest number of vacant houses per plantation reporting.

For all areas an average of 34 idle tillable acres was reported for each vacant house. The Arkansas River with the highest frequency of vacant houses on plantations had the low average of 14 idle tillable acres per vacant house. Even if no land were allowed to go fallow, cotton plantations would have been able to house and provide crop land for less than 5 percent more families than were resident upon them at the time of the survey.[27]

---

[26]Only those houses were counted which could be made habitable by repairs of $50 or less.

[27]On the 646 plantations in 1934 there were 9,414 tenant and laborer families (Appendix B, Table E), and 491 vacant houses (Table 57). There was probably no idle land on a few of the plantations with vacant houses.

DISPLACEMENT OF TENANTS AND RELIEF

Displacement of tenants, principally croppers, is an important factor in the rural relief situation in certain localities in the Southeast. Shifts from tenancy into the ranks of casual employment have resulted in relegating large numbers of farm families to the relief level. In five counties in the Old South Cotton Area in October 1933 a large proportion of the heads of relief families usually engaged as farm operators had been displaced.[28] Of the heads of relief cases who reported farm owner as their usual occupation, 75 percent still retained that status, while 18 percent were unemployed. On the other hand, of

Table 57—PLANTATIONS WITH TILLABLE ACRES IDLE AND WITH VACANT HOUSES,[a] BY AREAS, 1934
(Cotton Plantation Enumeration)

Area	Total Plan-tations	Plantations Reporting Vacant Houses	Percent Reporting Vacant Houses	Number of Vacant Houses	Number of Vacant Houses per Plantation Reporting	Idle Acres on Plantations Reporting Vacant Houses	Idle Acres per Vacant House
Total	646	180	28	491	2.7	16,674	34
Atlantic Coast Plain	56	10	18	24	2.4	636	27
Upper Piedmont	40	9	23	11	1.2	343	31
Black Belt (A)[b]	112	32	29	61	1.9	1,847	30
Black Belt (B)[c]	99	35	35	91	2.6	3,945	43
Upper Delta	133	37	28	151	4.1	4,681	31
Lower Delta	50	14	28	27	1.9	1,784	66
Muscle Shoals	22	-	-	-	-	-	-
Interior Plain	30	7	23	10	1.4	1,230	123
Mississippi Bluffs	47	13	28	38	2.9	825	22
Red River	28	6	21	16	2.7	500	31
Arkansas River	29	17	59	62	3.6	883	14

[a]Enumerators were instructed to enter only those vacant houses "judged to be habitable, or which could be made so with $50.00 or less repair."
[b]Cropper and other share tenant majority.
[c]Renter majority.

relief families whose heads reported cropper as their usual occupation, only 40 percent were so engaged in October 1933, while 53 percent were unemployed. Of those whose usual occupation was other tenant, 73 percent were still so engaged and 19 percent were unemployed.

In the Eastern Cotton Area[29] agricultural displacement had affected an even larger percentage of the cases on relief in June 1935 than was revealed in the October 1933 survey (Table 58). The proportion of displaced farm owners was smaller in June 1935 (16 percent), but the proportions of croppers and other tenants who were unemployed were higher, being 57 and 26 percent, respectively. Very few of the cases not employed in their usual or a higher agricultural status had other employment. Most of those who had obtained some other type of employment were

---

[28]McCormick, T. C., *op. cit.*, Tables 54A and B, pp. 106-107.

[29]Survey of Current Changes in the Rural Relief Population (32 sample counties), Division of Social Research, W.P.A.

still in agriculture, only an insignificant proportion having shifted to non-agricultural jobs.

A large 1930-1935 decrease in the number of tenants is a factor in the rural relief situation. This is indicated by a separate tabulation of current and usual occupation of the heads of June 1935 rural relief households in eight counties which showed in the 1935 Census of Agriculture a decrease of more than 10 percent in the number of all tenants (Table 58). When compared with the total 32 counties, these 8 selected counties had

Table 58—CURRENT AND USUAL OCCUPATION OF RURAL RELIEF HOUSEHOLDS IN THE EASTERN COTTON AREA, JUNE 1935
(32 Sample Counties)

Current Occupation (February–June 1935)	Usual Occupation (Last 10 Years)			
	All Farm Operators	Owners	Croppers	Other Tenants
	Total Sample			
Total: Number	2,170	458	1,066	646
Total: Percent	100.0	100.0	100.0	100.0
Employed in usual or higher agricultural status	56.2	82.1	36.8	70.0
Employed in lower agricultural status	3.3	1.3	4.3	3.1
Employed in non-agriculture	1.5	0.9	1.9	1.2
Unemployed	39.0	15.7	57.0	25.7
	Selected Counties[a]			
Total: Number	562	72	336	154
Total: Percent	100.0	100.0	100.0	100.0
Employed in usual or higher agricultural status	43.1	72.2	25.0	68.8
Employed in lower agricultural status	5.7	-	8.3	2.6
Employed in non-agriculture	0.3	-	0.6	-
Unemployed	50.9	27.8	66.1	28.6

[a]Eight selected counties in which the decrease in number of all tenants 1930-1935 was in excess of 10 percent as determined by the 1935 Census of Agriculture.

Source: Survey of Current Changes in the Rural Relief Population, Division of Social Research, W.P.A.

a noticeably smaller proportion of heads usually engaged as farm owners or croppers who were currently so employed, with a much larger proportion unemployed. Among other tenants the same situation existed but the difference was slight. In the selected counties twice as large a proportion of croppers had been demoted to the status of farm laborer. Evidently in those areas where there has been a relatively large decrease in the number of all tenants during recent years, displaced croppers constitute a larger portion of the relief load. In areas of heavy tenant displacement, loss of farm ownership appears to be relatively more frequent also.

A further idea of the importance of displacement of tenants as a factor in the relief situation may be obtained when it is noted that in Eastern Cotton Area counties in June 1935 unemployed croppers and other tenants accounted for 37 percent of the agricultural relief cases, or approximately 20,000 families in the whole area.[30] The 1934 study of nine counties in North Carolina, considered fairly representative of the State, exclusive of the Mountain region, revealed that displacement of tenants had been an important factor in the relief problem, though not as important as in the Eastern Cotton Area proper. In the counties included in this study 24 percent of agricultural relief cases were displaced croppers or displaced other tenants.[31] In a number of counties in eastern North Carolina displaced tenants comprised more than three-fourths of the agricultural relief cases in 1934. The conclusion is reached that, although tenant displacement may not have caused excessively high rural relief rates generally, it has been an important factor in the relief situation in some cotton areas.

RELIEF AMONG PLANTATION FAMILIES

The unique and in some ways feudalistic characteristics of the plantation system lead to especial considerations of relief needs and relief policies relative to families resident on plantations. Considerable area differences may be noted in the percentage of plantation families who have received relief[32] (Table 59). In the whole Eastern Cotton Belt it appears that 1 plantation family in every 5 received relief at sometime during the period January 1, 1933 to June 30, 1935, but the largest proportion of relief among plantation families seems to have been concentrated in 5 of the 11 areas covered in this study: Red River, Arkansas River, Mississippi Bluffs, Interior Plain, and Lower Delta. In these areas the proportion of plantation families who received relief at any time ranged from 29 percent in the Lower Delta to 50 percent in the Red River, the areas of heavy relief on plantations all being located in the three States of Arkansas, Louisiana, and Mississippi. Relief among plantation families seems to have been especially infrequent in the Upper Delta and the Atlantic Coast Plain where only 2 and 4 percent, respectively, had relief status. In the Muscle Shoals, Upper Piedmont, and 2 Black Belt areas only approximately 1 plantation family in every 10 had been on relief rolls.

---

[30] Data on file in the Division of Social Research, W.P.A. as obtained from the Survey of Current Changes in the Rural Relief Population.

[31] Blackwell, Gordon W., *op. cit.* Sample counties included in this tabulation were Alexander, Bertie, Columbus, Greene, Iredell, Onslow, Stokes, Tyrrell, and Washington. A total of .3,374 rural relief cases were included in the sample.

[32] At sometime during 1933, 1934, or 1935 to July 1.

The tenure status of plantation families seems to have a decided bearing upon relief status (Table 60).[33] As is to be expected, more of the displaced tenant families—those families living on the plantation with neither a crop nor a definite work

Table 59—RELIEF STATUS OF 5,147 PLANTATION FAMILIES, BY AREAS
(Cotton Plantation Enumeration)

Area	Total Families Reporting[a]		Received Relief[b]		Did Not Receive Relief	
	Number	Percent	Number	Percent	Number	Percent
Total	5,147	100.0	1,163	22.6	3,984	77.4
Atlantic Coast Plain	424	100.0	18	4.2	406	95.8
Upper Piedmont	773	100.0	65	8.4	708	91.6
Black Belt (A)[c]	687	100.0	81	11.8	606	88.2
Black Belt (B)[d]	248	100.0	25	10.1	223	89.9
Upper Delta	131	100.0	2	1.5	129	98.5
Lower Delta	357	100.0	103	28.9	254	71.1
Muscle Shoals	397	100.0	31	7.8	366	92.2
Interior Plain	1,332	100.0	486	36.5	846	63.5
Mississippi Bluffs	361	100.0	149	41.3	212	58.7
Red River	212	100.0	106	50.0	106	50.0
Arkansas River	225	100.0	97	43.1	128	56.9

[a]Data not available for 24 cases.
[b]Received Federal relief at some time during the period January 1, 1933 through June 30, 1935.
[c]Cropper and other share tenant majority.
[d]Renter majority.

agreement—had received relief (43 percent) than was the case with families with ordinary tenant or laborer status. Croppers and other share tenants had been on relief much more frequently than renters or wage hands, probably because relief was needed

Table 60—RELIEF STATUS OF 5,171 PLANTATION FAMILIES, BY 1934 TENURE STATUS
(Cotton Plantation Enumeration)

Tenure Status 1934[a]	Percent Received Relief			
	During 1933, or 1934, or 1935 to July 1	During 1933	During 1934	During 1935 to July 1
All families	22.6	19.1	14.3	3.1
Renters	13.3	8.3	8.3	2.0
Other share tenants	34.2	27.0	25.2	2.1
Croppers	24.9	17.6	14.4	3.7
Wage hands	11.5	8.0	8.4	1.3
Displaced tenants	42.6	14.9	34.0	25.5

[a]Ninety-four percent of the families were in the same tenure status in 1933 as in 1934; the tenure status for 1935 was not available.

primarily where the furnishing system ceased to function. One other share tenant in every three had received relief at one time or another, while this was true of only one wage hand in every eight. An analysis of yearly relief rates among plantation

[33]Tenure status is for 1934; 94 percent of the families were in the same tenure status in 1933 as 1934; the tenure status for 1935 was not available.

families reveals that almost as many were dependent at least in part upon relief in 1934 as in 1933. Except in the case of the few displaced tenant families, almost no relief was given among plantation families during the first half of 1935.

Monthly relief rates among these plantation families during 1934 and the first 6 months of 1935 indicate that after a peak in February 1934 the number on relief among all tenure groups except displaced tenant families declined rapidly and steadily to June 1935 with almost no seasonal variations from the trend (Appendix Table 62 and Figure 37). The fact that the decrease

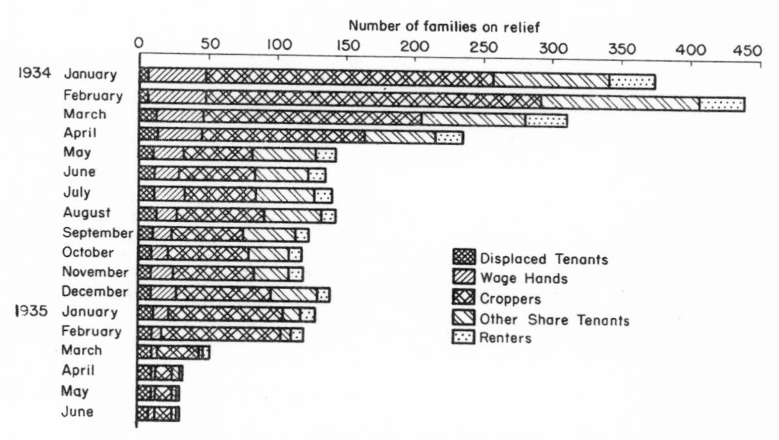

FIG. 37 - INCIDENCE OF RELIEF AMONG 5,033 PLANTATION FAMILIES, BY TENURE STATUS
January 1934 through June 1935

Source: Appendix table 62

AF-1417, W.P.A.

in the relief rate among plantation families between January 1934 and June 1935 (7.4 percent to 0.6 percent) was much greater than was the decrease in the combined relief and rehabilitation rate in the Eastern Cotton Area during the same period (17.1 percent to 10.7 percent) indicates the effect of new administrative policies. These were designed to end the practice among some plantation operators of shifting to the relief administration the burden of advancing to tenants short term credit for living expenses. Also, landlords doubtless were enabled by better cotton prices again to assume their usual responsibilities for advancing food. It should be noted that, aside from minor fluctuations, the number of displaced tenant families receiving relief remained fairly constant.

Moreover, a rather heavy turnover in the plantation relief population is suggested by these monthly relief rates. In only 2 months in 1934, January and February, was the relief rate as much as half that for the entire year. Although the relief rate was declining throughout 1934, it appears that new families were coming on relief rolls in significant numbers, replacing old clients.

Table 61—AVERAGE AGE OF HEADS OF RELIEF AND NON-RELIEF PLANTATION FAMILIES, BY TENURE STATUS, 1934
(Cotton Plantation Enumeration)

Tenure Status	Total		Received Relief[a]		Did Not Receive Relief	
	Number of Families[b]	Average Age of Head	Number of Families	Average Age of Head	Number of Families	Average Age of Head
Total	5,142	41.2	1,158	43.5	3,960	40.6
Wage hands	854	33.1	98	35.5	755	32.8
Croppers	2,872	41.0	709	41.9	2,144	40.6
Other share tenants	716	46.7	244	48.2	469	45.9
Renters	653	46.6	87	50.6	565	46.0
Displaced tenants	47	50.4	20	52.7	27	48.7

[a]Received Federal relief during 1933, 1934, or 1935 to July 1.
[b]Includes 24 families whose relief status was not available.

Contrary to the situation generally found throughout rural areas of the Eastern Cotton Belt, proportionately more Negroes than whites on plantations had received relief, the percentages being approximately 23 and 18, respectively.

Table 62—SIZE OF PLANTATION FAMILIES AND EMPLOYABILITY OF MEMBERS,[a] BY RELIEF STATUS AND BY 1934 TENURE STATUS
(Cotton Plantation Enumeration)

1934 Tenure Status	Total			Received Relief[b]			Did Not Receive Relief		
	Number of Families[c]	Number of Persons per Family	Number of Employable Persons per Family	Number of Families	Number of Persons per Family	Number of Employable Persons per Family	Number of Families	Number of Persons per Family	Number of Employable Persons per Family
Total	5,159	4.2	2.5	1,161	4.4	2.5	3,974	4.1	2.4
Renters	651	5.0	2.8	87	5.1	3.0	563	5.0	2.8
Other share tenants	715	4.5	2.7	243	4.5	2.6	469	4.5	2.7
Croppers	2,880	4.4	2.5	709	4.5	2.5	2,152	4.3	2.5
Wage hands	864	2.9	1.9	102	3.4	2.0	761	2.8	1.8
Displaced tenants	49	3.1	1.4	20	2.9	1.6	29	3.2	1.3

[a]Sixteen years of age and over and able to help with farm work.
[b]Received Federal relief during 1933, 1934, or 1935 to July 1.
[c]Includes 24 families whose relief status was not available.

The average age of the heads of families who had received relief was regularly higher than that of non-relief families, regardless of tenure status (Table 61). The average age of heads of wage hand families was considerably less than that of heads in any tenant group. Other share tenants and renters were much older than croppers. The displaced tenant families had still older heads.

Plantation families who had received relief tended to be larger than those with no relief status, the average size of family being 4.4 and 4.1, respectively (Table 62). This difference is noticeable to a significant degree only among wage hand and cropper families, while among displaced tenant families the non-relief group shows larger family units. Evidently a large family, usually considered one of the cropper's greatest assets, did not tend to keep plantation families off relief. Indeed with acreage reduction a large family may now be a liability to a cotton tenant. Scarcity of employable members in the household did not seem to account even partially for the need of relief assistance, as relief families on plantations had as large an average number of employable members as non-relief families, 2.5 as compared with 2.4.

Only 1 family in every 100 enumerated was classified as a displaced tenant family, that is, a family without a definite crop or work agreement with the landlord. Of the 49 displaced tenant families interviewed on the 646 plantations, 33 contained no employable member and 16 had able-bodied members. Families with widowed or unmarried females as heads were found most frequently in the unemployable group, there being 18 such cases. The heads of 13 families were above 64 years of age with no person over 16 in the household able to help with farm work, and 2 family heads were physically disabled. Negroes were noticeably over-represented in the displaced family group.

It appears obvious that as tenants become unable to tend a crop they usually leave plantations, just as in industry workmen must look forward to the time of forced retirement. Some few widowed or aged tenants are allowed to continue on plantations, however, picking up occasional odd jobs and tending a small garden patch rent free. The fact that only 4 displaced tenant families were found among the more than 2,500 families interviewed in the 5 areas, where the largest and most commercialized plantations are found (Upper Delta, Mississippi Bluffs, Red River, Arkansas River, and Interior Plain), indicates that families are not allowed to remain on these plantations after becoming an economic liability.

COSTS OF RELIEF AND SIZE OF RELIEF BENEFITS

Per capita expenditures for emergency relief have been lower in general in the southern States than in other parts of the country (Figure 38), yet more than 287 million dollars were expended for emergency relief purposes in the seven southeastern cotton States during the 33-month period January 1933 through September 1935 (Appendix Table 63). This means an expenditure of $17.49 for every person living in these States in 1930. By States, the per capita amount varied from $7.25 in Alabama to $24.41 in Louisiana. The per capita figure for the seven southeastern cotton States was 56 percent of the per capita figure for the country as a whole

The South has relied to a larger extent upon Federal funds for handling the relief situation than has been the case in other parts of the country. More than 96 percent of the obligations incurred for emergency relief during the period January 1933 through September 1935 in the seven States considered in this study were met with Federal funds (Appendix Table 63), as compared with 72 percent of the obligations in the country as a whole. When the per capita amount of Federal funds is considered, differences between the southeastern States and the Nation are not so great. Especially in the matter of State participation has the South lagged behind. Of the seven States,

FIG. 38 - PER CAPITA AMOUNT OF OBLIGATIONS INCURRED FOR EMERGENCY RELIEF
January 1933 through September 1935

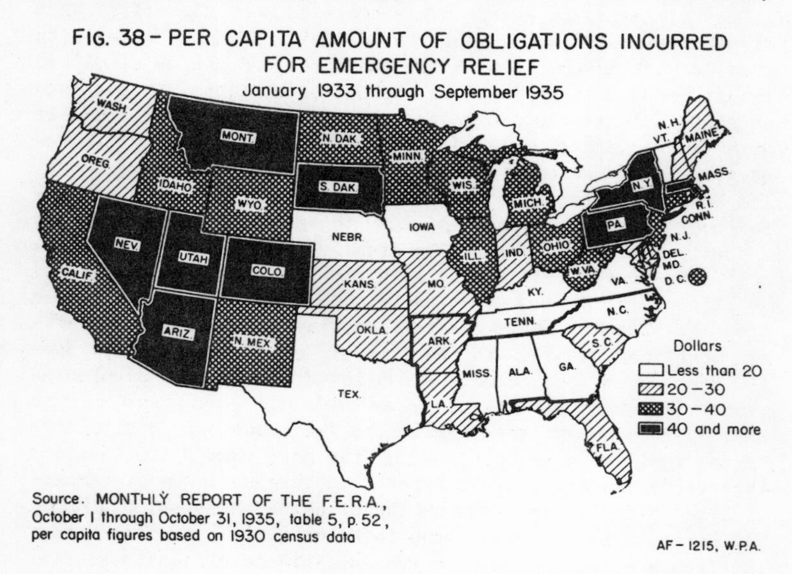

Source. MONTHLY REPORT OF THE F.E.R.A., October 1 through October 31, 1935, table 5, p. 52, per capita figures based on 1930 census data

AF – 1215, W.P.A.

only three—Arkansas, Alabama, and Mississippi—have made appreciable contributions and even here State funds amounted to less than 1 percent of the total.

Just as the southeastern cotton States expended a much smaller per capita amount for emergency relief, so too was the size of relief benefits per family assisted comparatively low (Table 63). From the beginning of the relief program there was a significant increase in the size of relief grants per family assisted in areas outside of principal cities throughout the country. In four of the seven southeastern cotton States the proportionate increase was decidedly larger than the increase in the country as a whole. Yet the average size of relief grants in September 1935 in each of these States was from 22 to 60 percent less than the national average of $20.23 for all cases outside of principal cities. The average relief grant per case for September 1935 in the southeastern States ranged from $8.00 in South Carolina to $15.77 in Louisiana.

In the Southeast as in other areas average amounts expended for agricultural cases were lower than for non-agricultural ones (Appendix Table 64). The median size of relief grants was

Table 63—MONTHLY RELIEF BENEFIT PER FAMILY (OUTSIDE OF PRINCIPAL CITIES) IN UNITED STATES AND SEVEN SOUTHEASTERN COTTON STATES, SEPTEMBER 1933, 1934, AND 1935

State	September 1933	September 1934	September 1935	Percent Increase September 1933 to September 1935
United States	$12.69	$18.96	$20.23	59.4
Alabama	6.26	12.59	13.15	110.1
Arkansas	4.85	11.26	13.13	170.7
Georgia	8.59	10.32	12.17	41.7
Louisiana	13.47	17.77	15.77	17.1
Mississippi	5.52	10.14	11.49	108.2
North Carolina	6.19	9.31	11.50	85.8
South Carolina	7.50	9.67	8.00	6.7

Source: *Monthly Reports*, Division of Research, Statistics, and Finance, F.E.R.A.

$11.84 for agricultural cases and $14.64 for non-agricultural cases. While 90 percent of the agricultural relief cases in the Eastern Cotton Area received less than $20 during the month of June 1935, this was true for only 78 percent of those whose usual occupation was non-agricultural. The medians were $8.98 and $12.44, respectively. With the exception of the Western

Table 64—AVERAGE[a] AMOUNT OF RELIEF RECEIVED BY RURAL HOUSEHOLDS DURING JUNE 1935,[b] BY USUAL OCCUPATION OF THE HEAD, IN THE EASTERN COTTON AREA (32 Sample Counties)

Usual Occupation of Head	Median Amount of Relief		
	All Cases[c]	White	Negro
Agricultural cases	$ 8.98	$10.05	$7.00
Owners	9.47	9.58	8.88
Tenants	10.19	11.74	5.98
Croppers	10.09	10.49	8.69
Farm laborers	7.75	8.91	6.55
Non-agricultural cases	12.44	13.75	9.92

[a]Median.
[b]Exclusive of cases opened, reopened, or closed during the month.
[c]Based on 3,308 agricultural cases of whom 2,184 were white and 1,124 Negro; and upon 2,152 non-agricultural cases of whom 1,514 were white, and 638 Negro.
Source: Survey of Current Changes in the Rural Relief Population, Division of Social Research, W.P.A.

Cotton and Appalachian-Ozark Areas these medians were much lower than those for other areas. They doubtless reflect an effort to keep relief grants in line with the low wages prevailing in agriculture.

Significant differences may be noted in the size of relief grants in June 1935 in the Eastern Cotton Area when calculated by tenure status and color of head (Table 64). Farm tenants received larger grants on the average than farm owners. The average size of grants to farm laborers during the month of June—$7.75—was much smaller than that to other agricultural cases. In all groups Negroes received noticeably smaller grants than whites.

Various other studies have shown that rural Negroes in the Eastern Cotton Area have generally been under-represented on relief rolls, have received smaller relief benefits than whites and have been less likely to be assigned to work relief jobs.[34] This situation is to be expected in view of racial attitudes in the South. It is probable that relief officials accepted the low scale of living of the great majority of rural Negroes as the basis for determining budgetary deficiencies. This may have resulted in fewer Negroes being considered eligible for relief and for smaller grants being extended to those cases accepted for relief.[35]

The average relief grant figures given in Table 64 for rural cases by occupation of the head and by color must be qualified to some extent. The group of cases represented in the sample contained only cases which were not opened, reopened, or closed during the month. In these figures the size of family was not taken into account, nor could the extent to which relief was only supplementary to private income be considered.

It should not be assumed that these cases received a like amount 12 months in the year. Turnover in the rural relief population was rapid as closings were balanced by newly opened and reopened cases. This fact was made evident above where it was shown that although 14 percent of the plantation families had received relief in 1934, less than 5 percent received assistance in any single month, with three exceptions (Appendix Table 62). A study of a sample of rural and town families on relief rolls during the period May through October 1934 revealed that less than one-third of those in the Eastern Cotton Area received relief during all 6 of the months.[36] It is therefore evident that the great majority of rural cases were dependent upon relief for only a part of each year.

Comparison is sometimes made between the size of relief grants to tenant families and the usual amount of landlord subsistence

---

[34] Mangus, A. R., *op. cit.*; Mangus, A. R., *Type and Value of Relief Received by Rural and Town Cases, October 1934*, Research Bulletin F-8, Division of Research, Statistics, and Finance, F.E.R.A., April 24, 1935; Beck, P. G. and Forster, M. C., *op. cit.*, pp. 31-37; and McCormick, T. C., *op. cit.*, Table 2, p. 79.

[35] Mangus, A. R., *The Rural Negro on Relief, February 1935*, Research Bulletin H-3, Division of Research, Statistics, and Finance, F.E.R.A., October 17, 1935, p. 1.

[36] Mangus, A. R., *Relief History, May to October 1934, of Rural and Town Relief Cases*, Research Bulletin F-9, Division of Research, Statistics, and Finance, F.E.R.A., April 24, 1935.

advances. In such a comparison average yearly relief grants should be considered as well as average monthly relief grant figures. Plantation families on relief in 1934 received an average of less than $27 during the entire year (Table 65). This amounted to an average monthly grant of $7.70 during the 3½ months they were on relief. Displaced tenant families received by far the largest amount of relief, their average yearly grant amounting to almost $61. Renters received the next largest relief grants and other share tenants the lowest, the average yearly amounts being approximately $32 and $21, respectively. Cropper and wage hand families received almost the same amounts of assistance, averaging $24 and $25, respectively, for the entire year.

Table 65—AVERAGE ANNUAL AND MONTHLY RELIEF GRANTS OF 290 PLANTATION FAMILIES, AND NUMBER OF MONTHS ON RELIEF, BY TENURE STATUS, 1934
(Cotton Plantation Enumeration)

Tenure Status	Total Families Reporting Amount of Relief	Amount of Relief per Family 1934	Average Number of Months on Relief	Average Monthly Relief Grant 1934
Total	290	$26.95	3.5	$7.70
Renters	52	31.90	3.9	8.18
Other share tenants	34	21.09	3.5	6.03
Croppers	155	23.97	3.0	7.99
Wage hands	35	24.94	4.0	6.24
Displaced tenants	14	60.79	7.3	8.33

Comparison of the average amount of relief received by these plantation families in 1934 with the average amount of subsistence advances from landlords reveals that relief assistance has been relatively small. The tenants for whom relief data are available received an average of approximately $7.70 per month from relief, while tenants being furnished by landlords received subsistence advances averaging $12.90 per month.

THE RURAL REHABILITATION PROGRAM

In the spring of 1934 the rural rehabilitation program was inaugurated under the Federal Emergency Relief Administration as one step toward a differentiated treatment of the numerous relief groups and relief problems.[37] Its objective was to remove farm families from relief rolls by advancing credit for subsistence and farming operations so that they could again become self-supporting. It was assumed that families on farms would be able to produce a major part of their own food if land,

[37]For a more detailed discussion of the rural rehabilitation program under the F.E.R.A. see *Monthly Report of the F.E.R.A.*, August 1935, pp. 14-24; also Asch, Berta, *Farm Families on Relief and Rehabilitation*, forthcoming monograph, Division of Social Research, W.P.A.

work stock, equipment, and short term credit were made available to them. Up to that time, the Federal government had not provided this type of credit to tenants in the South.[38]

As the program was worked out, emphasis was placed primarily upon subsistence farming, although in money crop areas commercial farming on a family basis often was allowed. Formulation of the details of the program was left largely to the discretion of the individual States, this policy being necessitated by major regional differences in type of farming, size of farms, and the tenure status of farm families on relief. To enable the large majority of cases to repay the short term loans for current operating and subsistence expenses and to pay the yearly installments on capital goods purchased, it was necessary to provide an opportunity for most clients to make payments through labor on rehabilitation and E.R.A. work projects.

In the main, the program in 1934 was confined to families already living on farms, although some village and town cases with farming experience were moved to rural areas. As the program ran into its second crop year, its functions began to broaden beyond rehabilitation of farm families "in place." State Rehabilitation Corporations of a non-profit and self-liquidating type were formed in most of the States, these agencies receiving grants from the F.E.R.A. and having the rights of private corporations, including the right of buying, selling, and leasing land, buildings, and capital goods. Debt adjustment efforts resulted in scaling down debts of many farm owners. Plans for the creation of organized farm communities, work centers, commodity exchanges, and cooperatives were formulated and a few were set up in 1935. Occasional dormant industrial plants, quarries, and the like, usually employing part-time farmers, were resuscitated and seasonal industries were developed in certain localities. Resettlement of farm families then on submarginal land or of families in stranded rural industrial communities was attempted on a small scale.

On June 30, 1935, the entire program was transferred to the State Rehabilitation Corporations, which in most States voted to transfer their activities to the Resettlement Administration through which future funds were to be made available.

This type of program, particularly the phase of rehabilitating farm families "in place", was especially suitable to the southern cotton area. In effect what it offered for most of the families, other than farm owners or some few tenants who were aided in buying farms, was a substitution of government credit for landlord credit, where the latter was no longer available.

---

[38] See chapter V.

Credit from the existing governmental agencies could not, under their rulings, be extended to such poor risks as former tenants on relief. The reasons why the Rehabilitation Corporations could afford to do this were twofold: (1) the credit extended did not greatly exceed the amount which would have been necessary to feed the family on the dole; (2) a farm plan was worked out for the family and the existing plan was supervised.

Beginning with a small number of clients in the spring of 1934, the rural rehabilitation program expanded slowly during the first year and more rapidly in 1935. Immediately prior to its transference to the Resettlement Administration on June 30, 1935, it was making advances to more than 200,000 cases in 43 States (Appendix Table 65).

Approximately 138,000 cases in the 7 southeastern cotton States were under the care of the program at one time or another. There was wide variation in the extent of the program in the various States, Louisiana reporting 37,000 cases and North and South Carolina only 8,000 and 7,000, respectively. Among other factors, three may be mentioned which were partially responsible for the extent of development of the program in a given State: (1) the type of program formulated by the State rural rehabilitation division; (2) the number of farm relief cases considered to be capable of rehabilitation; and (3) the availability of crop land and houses in the areas where dispossessed farm families were located.

Small cotton farmers were particularly in need of such a program. To the thousands of displaced tenants whose re-entry into the agricultural economy had been effectively blocked by the acreage control program, it offered the only opportunity to become even partially self-supporting. The program answered a real need in this area and developed relatively rapidly.

In plantation areas with large-scale and highly capitalized commercial farming, however, the program took root only gradually. On the 646 plantations studied, there was only one rural rehabilitation client in 1934, although 719 of the 5,033 plantation families reporting received relief during the year (Appendix Table 62). Evidently some plantation owners were willing to be subsidized by relief payments to their tenant and wage hand families, but at the same time they refused to cooperate in the rural rehabilitation program. Relatively little idle crop land and few vacant houses, which the program called for rent free or at regular rental terms, were made available by planters. With profitable commercial farming the objective, and with cotton prices rising, they had little idle land other than that which soil conservation policy required to go fallow. Furthermore, if a resident family on a plantation was capable of farming, the planter undoubtedly preferred to make his own farming agreement with the tenant rather than work through a government agency. If he could shift the burden of furnishing the tenant to the relief administration, he was so much to the good. In plantation areas the Rehabilitation Corporation had to seek available land and houses among small farm owners or on plantations in the hands of land banks or mortgage companies.

Goods to the amount of almost 50 million dollars were issued to rehabilitation clients during the 15 months the program was in operation under the F.E.R.A. (Appendix Table 65). Of this amount 29 percent was for subsistence goods, such as food, clothing, and other items of living, and the balance for farm supplies and capital goods, including seed, fertilizer, household and farming equipment, and work stock. In addition, obligations totaling 11 million dollars were incurred for purchases of materials, particularly for the construction of family dwellings for clients, and for administrative expenses.[39]

In the southeastern cotton States, the amount of all goods issued per case, when figured for the total number of clients ever on the program, was much higher than that for the whole country, the average being $166 as compared with $124.[40] This difference is more noticeable for subsistence than for rehabilitation goods, and probably is due to the fact that clients remained on the program longer in the South than in other parts of the country.

The amount of goods issued per case varied greatly from State to State in this area, according to the type of program formulated and the number of cases accepted for loans. Louisiana and Alabama, with per case averages of $87 and $105, respectively, for the value of all goods issued, had the smallest loans per case and the largest number of cases assisted. In these two States the program was mainly of a subsistence type with relatively few loans for capital goods. On the other hand, Georgia and South Carolina each made capital goods loans averaging more than $220 per case.

Loans to clients receiving advances in June 1935 in the Eastern Cotton Belt, exclusive of non-cotton areas, averaged $175 for all goods issued since the date of their enrollment (Table 66). The average is higher than the $166 for all cases ever on the program in the seven States considered as a whole (Appendix Table 65). This may be explained by the fact that the cases included in the Eastern Cotton Area must have received advances

---

[39] *Monthly Report* of the F.E.R.A., August 1935, p. 20.

[40] In considering these figures it should be kept in mind that many cases were on the program for a short period of time and received only small loans, while others on the program during the growing season of the greater part of 2 crop years received relatively large advances. Also, many families were accepted for rehabilitation when only subsistence advances were needed while others required large capital outlays to enable them to conduct farming operations efficiently. For these reasons the spread from the average is large.

in June 1935 and therefore may have been on the program during most of the growing season of 2 crop years, while among all cases ever on the program there is more likelihood that some were under care for only a few months or only in 1934. Also

Table 66—AVERAGE AMOUNT OF GOODS ISSUED SINCE DATE OF ENROLLMENT TO CASES RECEIVING RURAL REHABILITATION ADVANCES IN JUNE 1935 IN THE EASTERN COTTON AREA, BY TYPE OF GOODS AND BY COLOR
(32 Sample Counties)

Color	Total Cases Represented	Amount of All Advances per Case	Advances for Subsistence Goods		Advances for Capital Goods	
			Percent of All Cases	Amount per Case	Percent of All Cases	Amount per Case
Total	6,288	$175	98.2	$69	90.6	$119
White	4,028	205	99.2	74	91.5	145
Negro	2,260	122	96.6	60	89.0	73

Source: Survey of Current Changes in the Rural Relief Population, Division of Social Research, W.P.A.

it is probable that cases in the commercial farming cotton area required larger advances than those in subsistence farming areas.

While 98 percent of the cases received subsistence advances, only 91 percent received loans for rehabilitation or capital goods (Table 66). Negro-white differences in the amount of goods issued followed those usually found for size of relief

Table 67—USUAL AND CURRENT OCCUPATION[a] OF HEADS OF RURAL CASES RECEIVING RELIEF AND UNDER CARE OF RURAL REHABILITATION PROGRAM, IN THE EASTERN COTTON AREA, JUNE 1935
(32 Sample Counties)

Occupation	Relief Cases		Rehabilitation Cases	
	Usual	Current	Usual	Current
Total: Number	6,356	6,356	6,170	6,170
Total: Percent	100.0	100.0	100.0	100.0
Agriculture	57.8	25.3	92.2	94.7
Farm operators	34.2	20.7	82.1	92.9
Owners	7.2	6.5	18.7	20.4
Tenants	10.2	7.7	36.6	56.5
Croppers	16.8	6.5	26.8	16.0
Farm laborers	23.6	4.6	10.1	1.8
Non-agriculture	38.0	7.2	7.8	0.3
No usual occupation or currently unemployed	4.2	67.5		5.0

[a]Usual occupation is defined as the occupation at which the head was engaged most during the last 10 years; current occupation is the one at which the head had been engaged for at least one full week between February 1 and June 30,1935.

Source: Survey of Current Changes in the Rural Relief Population, Division of Social Research, W.P.A.

benefits among rural families in the South, the average of $122 in all goods for Negro cases being decidedly lower than that of $205 for whites. The difference was much greater for capital goods than for subsistence advances. Likewise a somewhat smaller proportion of Negro than white cases received subsistence goods and the same was true for capital goods. Evidently fewer Negro cases received both types of goods than did whites.

To understand clearly the nature of the rural rehabilitation program in the Eastern Cotton Belt and to know how it was applied, it is necessary to note what types of families were accepted for rehabilitation. An analysis of the usual occupation of relief and rehabilitation cases in June 1935 reveals that, as would be expected from the nature of the program, only 8 percent of the rehabilitation clients had usually been engaged in non-agricultural industry during the past 10 years, whereas 38 percent of those remaining on relief rolls were non-agricultural cases (Table 67). Likewise only a small proportion of the rehabilitation group represented farm laborers. More than four-fifths of the rehabilitation cases in this area came from among the farm operators. Roughly speaking, of every nine of these farm operator cases two were farm owners, three were croppers, and four were other tenants. Thus, those farm families who were presumably in a better financial situation (owners, share tenants, and renters) were more often taken on the rehabilitation program than were croppers and farm laborers.

Study of current tenure status of relief and rehabilitation cases indicates that the rehabilitation program at least temporarily benefited families taken under its care (Table 67). Eleven percent more of the rehabilitation cases were currently in the farm operator group than were usually in that group. The number of farm owners had increased slightly and evidently a large number of croppers and laborers had shifted up the agricultural ladder. While the proportion usually engaged as croppers (27 percent) decreased to 16 percent currently so employed, and laborers from 10 percent to 2 percent, the proportion among other tenants increased from the usual occupation ratio of 37 percent to a current figure of 57 percent.

The rural rehabilitation program in the Eastern Cotton Belt varied considerably with the tenure status of the clients. By providing credit facilities, the program sought to enable farm owners to carry on farming operations efficiently, in some instances preventing mortgage foreclosures. Debt adjustments saved other owners from losing their farms. Some few tenants were enabled to make a down payment on a place and thus enter the farm ownership group. The program resulted in many clients growing more of their own food than formerly.

In general, the future outlook of the cotton tenant or laborer family on rural rehabilitation in some ways was not greatly unlike that of families under the often criticized sharecropper system. Tenure arrangements for a longer period than 1 year were seldom made, and farming practices were supervised. Clients received subsistence advances of about the same amount as landlords usually provided, and short term credit for fertilizer, seed, feed for work stock, and like operating expenses. Furthermore, a share of the crop or a specified amount of cash rent was usually paid for the use of the land either directly to the owner or to the Rehabilitation Corporation which in turn was renting from the owner.

The program differed from the usual operation of the tenant system in that there were no high interest rates and as a rule work stock and equipment were sold to the client with 3 to 5 years allowed for payment. Thus, the former wage hand or cropper perhaps had a somewhat better opportunity to improve his economic situation under this program than under the tenant system. Furthermore, diversification of crops was not only possible but obligatory under rehabilitation regulations in southern States. The government, assuming the responsibilities of the landlord, was in a position to dictate what should be planted by rehabilitation clients. Trained home economists were employed under the rehabilitation program to aid housewives with homemaking and family problems. Several years are required before the success or failure of such a program can be judged, but it appears evident to one familiar with its operation that the restorative aspects of the program were undoubtedly preferable to a continuation of relief with its merely palliative character.

In planning a long time rehabilitation program, the important factor to be considered is the type and capability of the families to be dealt with. Data on rehabilitation prospects among farm families on relief in the Southeast are scattered and should not be relied upon to too large an extent. A study of approximately 30,000 relief cases applying for rehabilitation in Alabama in 1934 revealed that 54 percent were accepted as capable of managing capital goods.[41] Thirty-five percent more were accepted to be advanced chiefly subsistence goods, and 11 percent were disapproved. The proportion of Negroes accepted as capable of managing capital goods was slightly larger than that of whites. Also, fewer Negro than white cases were subsequently cancelled.

A similar study of more than 20,000 cases in Arkansas in 1935 indicated that a larger proportion of the applicants, 77 percent, were accepted as capable of managing capital goods.[42] Of these one in every four was classified as a "best risk." Fourteen percent were accepted to receive subsistence loans only, while

---

[41] The study was directed by Harold Hoffsommer, then F.E.R.A. State Supervisor of Rural Research in Alabama, and later by John H. McClure, W.P.A. Temporary State Supervisor of Rural Research. No material as yet has been published from this study. Analysis was made of all applications for rehabilitation in Alabama in 1934. Classification of a family had been determined previously by relief officials and the local county rehabilitation committee, after a consideration of the composition of the client's household and his past history, together with various opinions concerning him.

[42] This study was directed by C. O. Brannen, W.P.A. State Supervisor of Rural Research in Arkansas. Classification of families was determined in the same way as in Alabama.

nine percent were disapproved. Negroes were decidedly underrepresented on rehabilitation rolls, the Negro ratio in the rural relief population being much higher than that among families on rural rehabilitation. Furthermore, fewer Negro than white applicants were accepted as capable of managing capital goods.

The capability of rural relief families for rehabilitation[43] was analyzed in a study made in 11 sample counties in North Carolina in 1934. It was found that because of unemployability 47 percent were unable to be self-supporting as farmers and an additional 5 percent had had insufficient farming experience. The remaining 48 percent appeared capable of rural rehabilitation from the point of view of available labor in the family, health, and farming experience. It is this latter group which corresponds fairly closely to the rehabilitation applicants studied in Alabama and Arkansas. Of the 1,854 such cases included in the North Carolina sample, 28 percent were rated as good prospects and 43 percent as fair prospects. There is some doubt as to how many of the fair prospects could be successful in a rehabilitation scheme even with fairly close supervision, but it is believed that most of the relief families in these first two groups, representing almost three-fourths of the employables with farming experience, were capable of managing capital goods. The remaining 29 percent were rated as bad risks. Of the tenants and laborers, only 140, or 11 percent, were deemed capable of eventually becoming successful farm owners if given an opportunity. It is significant that capability of families was rated lowest in cotton and tobacco counties with high tenancy rates.

Although the results of these various State studies of the possibilities of rural rehabilitation are not directly comparable, certain broad generalizations may be made from them. It is evident that the principle of rural rehabilitation cannot be dismissed as impractical due to lack of capability among rural relief families to be dealt with in the South. On the other hand, the idea that it is possible to enable every able-bodied farm family on relief to own and successfully operate "a mule and 20 acres" is equally unrealistic. Under the rural rehabilitation program of 1934-1935 with its supervisory facilities, it appears certain that more than one-half of the able-bodied farm families on relief in the southeastern States

---

[43] Blackwell, Gorden W., *Rural Relief Families in North Carolina*, North Carolina Emergency Relief Administration, Raleigh, North Carolina, 1935 (mimeographed). Rating of families was based on a consideration of family composition, farming history, and opinions of landlords or employers, social service workers, work project foremen, and E.R.A. farm and garden supervisors. Figures have been revised slightly since issuance of the report.

were capable of participating to the extent of attaining ownership of stock and equipment.

When one considers the possibilities of replacing the firmly entrenched tenant system with a system of small owners, the outlook is probably brighter among tenants who have not been relegated to relief status, as it is to be expected that in general the most capable farm tenants have continued to be self-supporting. Tenants with past relief status, however, make up only a minor segment, perhaps a fourth or a third, of the total tenant population in the Eastern Cotton Belt. Among those who have managed to remain independent of Federal relief, opportunities for increasing farm ownership with the aid of the government may be more encouraging.

Chapter XI

CONSTRUCTIVE MEASURES

The policy of the South and of the Nation toward the tenant system and cotton economy was, up to 1932, one of *laissez faire* with occasional mild efforts to promote diversification of crops or to provide palliatives such as cooperative marketing and the extension of government credit to those whose security was adequate. The results of this policy are apparent enough in the preceding chapters. First the slave plantation, then the tenant plantation, were based on an exploitative culture of a money crop with ignorant labor working under a paternalistic system, producing a raw material and exchanging it for finished products manufactured elsewhere. The process has been marked by increasing soil exhaustion, soil erosion, ruinous credit conditions, and mounting tenancy.

Those who would abandon the *laissez faire* policy in favor of promotion of constructive measures will, of necessity, have to face certain basic realities of the social and economic organization of the South and shape their plans accordingly. Otherwise they can expect conflict of wishful thinking with stubborn facts.

The effectiveness of specific constructive programs is conditioned by the quantity and quality of the population, the inter-regional and international implications of cotton economy, the type of farm organization to be promoted, and the time element necessary in effecting social change.

BASIC REALITIES

The People

The first realities are those which concern the number and capacity of the people of the Southeast. The Cotton Belt has for generations produced more people than were needed to cultivate the cash crop. These people reared on cotton farms found outlet for their productive capacity either in urban industry or elsewhere in the agrarian system of the Nation. A similarly large younger generation is now maturing and entering the labor market. Mechanized, commercial farming provides no certain future for these teeming thousands of maturing youth. In his summary of the population outlook of the Southeast,

Vance expresses the situation as follows:

> ...the region's problem of low standards can, in the long run, be met only by the adjustment of (1) a decreasing rate of population growth to (2) an increasing utilization of regional resources with (3) redistribution of part of the population. The foregoing analysis indicates doubt as to whether any one of these three factors is at present operating to the benefit of the area. Furthermore, any change in the birthrate of the population or any improvement in the economic life of the South is bound to take place very slowly. Migration remains the area's sole immediate recourse to soften the blow of lost markets or lift it from stabilized poverty in case of a return of pre-depression conditions. Even if a program of regional reconstruction larger than any yet contemplated could be carried swiftly and effectively to conclusion, the nation and the region should plan for migration. If world markets are lost, it will be desirable, if not absolutely necessary, that six or seven million persons should migrate from the Southeast....
>
> Certainly with seven millions wholly or partially removed from the consumers' market and pressing on the national labor market it is hardly reasonable to envisage industrial recovery. The maximum figure, then, does not measure migration so much as the amount of regional poverty that might be expected to prevail. The point can easily be made that these figures can be put to their most realistic use as an estimate of the relief burden likely to fall upon the nation should the South lose the whole export cotton market.[1]

While Vance has diagnosed the case with accuracy, the facts do not fully agree with the remedy he suggests (wholesale migration, or wholesale relief), for there is no assurance from the agricultural outlook, as described in other sections of Goodrich's analysis, that these six or seven million people could be absorbed into any better situation outside the South.[2] If they remain where they are, they at least have the advantages

---

[1]Goodrich, Carter, and Others, *Migration and Economic Opportunity*, Philadelphia: University of Pennsylvania Press, 1936, pp. 162-163.

[2]*Idem.*

of a salubrious climate where long growing seasons and abundant rainfall make life somewhat easier than in harsher climates. Nor do the data in this study support the theory that to remain and endeavor to work out a different agricultural economy from the one at present practiced would necessarily mean greater poverty than would be the case if the migrants join the millions who exist on the margin of the industrial labor market.

Not only must the future of the South be molded about a population policy which will fit the essential trends in numbers but it must also be adapted to the capacity of the people.

This means, first, that constructive effort must recognize the inter-racial character of the southeastern population and the peculiar traditional relationships between the races. This, however, is of decreasing importance as with the increase of white tenancy it becomes more and more a class rather than a race problem.

The comparative poverty of educational opportunity, especially for Negroes, provides an index of the lack of development of native capacities. The system of strict management of tenant operations and supervision of tenant expenditures further saps initiative and deadens the sense of self-reliance.

Wide differences exist in the educational attainment of farmers of the South and farmers of other sections. A still greater disparity is evident between the rural and the urban sections of the country. This disparity is due in part to the inferior school facilities and limited grade attainment of rural pupils but also in part to the fact that rural schools are to a great extent educating young people for the city. Such studies as have been made of the quality of migration indicate that the grade attainment of those moving from the farms is higher than that of those remaining. To produce an educated rural population in the future, therefore, would require that more of the rural pupils remain on the farm after finishing school.

This situation points to the need for intensified efforts for adult education or the training of farmers on the job and supervision of farming operations. The training and supervision needed is of the type given in the past by farm and home demonstration agents and rehabilitation supervisors to owner operators and independent tenants. Increase in the efficiency of production, introduction of specialty crops, and improved practices of animal husbandry have been principal objectives of these programs. Owing to the organization of the plantation system, however, it has been difficult for extension agents to reach tenants directly. Land use, crop practices, and animal breeding are planned and managed by the landlord. Any modification of these plans and practices is subject to his rulings. A cooperative landlord may, therefore, allow his tenants to receive extension instruction, while one who is not interested will block any constructive program proposed.

The success or failure of all of the constructive measures is more or less dependent on the intelligence and information of the people involved. It may be said, therefore, that the keystone of any program of southern agrarian reconstruction must be a program of mass education.

To suggest that the South might exert itself more in the direction of education is pertinent but somewhat in the nature of a recommendation that the region lift itself by its own bootstraps. In proportion to its wealth base the South is spending as much for education as any other section. However, the wealth base is relatively so small that the resulting per capita expenditures are far below those in other sections.

This situation points to the need for a Federal program of grants-in-aid for the equalization of educational opportunity, justified by the fact that the surplus wealth of the Southeast is concentrated in other sections and the further fact that such a large proportion of youth educated in the rural areas make their life contribution in industrial cities.

Inter-regional and International Relationships

A realistic program will also consider the fact that one of the basic features of the international trade of the United States has been the exchange of raw cotton and tobacco for the finished products of other sections and of other nations. The whole freight rate structure of the region is based on that hypothesis and tends to crystallize the situation. The reduction of cotton exports drastically alters the trade balance, but it is becoming increasingly apparent that if cotton is to hold an important position in the export trade of the United States, the cotton producers of the Southeast must compete with the farmers of South Africa, India, South America, Manchuria, and other potential cotton producing areas having low standards of living. From this it follows that the more cotton that is produced in other areas, the more the farmers of the Southeast must continue to produce in competition with families that have a low standard of living, thereby depressing the southern standard unless cotton is merely an incidental portion of farm income instead of the dominant factor which it now is. Thus, southeastern economy is dependent on the balance among regions and among nations, and changes in the present system must be weighed against resulting shifts in inter-regional and international relationships.

Large vs. Small Scale Operations

It should be pointed out that all the constructive measures discussed in the following section are predicated on a trend toward small farms cultivated by independent operators. However, there is a possibility that under a *laissez faire* policy and with increasing expansion of commercial farming and mechanization, a trend toward large scale farming may prevail. One school of thought visualizes increasing concentration of acreage into large landholdings operated by increasingly mechanized processes, the other, an increasing subdivision of land into family-sized farms on which money crops provide a minimum proportion of the income and subsistence crops balance the family budget. These conflicting philosophies are summed up in *Southern Regions* as follows:

> There were as usual contradictory pictures of the future of agriculture. One was a picture of the new agriculture as predominately machine farming, on large farms owned by commercial concerns with ever increasing use of inventions and the employment of fewer men. There would be the cotton picker which would do the work of forty Negroes or the multiple purpose corn harrow or threshing machine which might do the work of a hundred men. It was pointed out that, already by 1930, over half of all farms had automobiles, about 15 percent trucks, besides a great many other types of mechanical farm equipment. On the other hand, the picture was presented, following actual trends, to show that whenever the big farms owned by banks and insurance companies were sold, the tendency was invariably to break them up into smaller units. The statistics showed a regular decrease in the size of farms. Likewise, during the depression years the use of machine cultivation had decreased tremendously. The great decrease in exports threatened to make commercial farming unprofitable, while there seemed to be a definite trend toward self-sufficing farming, with a very large increase of the balanced live-at-home operations....[3]

It is not outside the realm of possibility that the two trends may develop simultaneously with larger and more mechanized farms for the production of money crops, interspersed with increasing numbers of family-sized farms which are operated primarily to produce a living for a family. These smaller farms must include small acreages of money crops to pay for taxes, clothing, and services, but in a region of as varied potentiality as the

---

[3] Odum, Howard W., Chapel Hill: University of North Carolina Press, 1936, p. 427.

South they can also produce the major portion of the family food requirements. Neither of these systems is highly efficient under a regime of share tenancy. Large-scale mechanized farming is most productive when managed by a central control and operated by wage laborers. Small-scale family farming is more efficient when conducted by owner operators or operators who at least own their work stock and finance themselves, provided they are capable farmers. Some of the advantages of large-scale operation may be secured by small farmers through closer cooperation.

The concentration of commercial farming on large mechanized acreages will mean the reduction of manpower required per acre and also of the total manpower employed on farms, unless the decrease in costs extends the market so that a far greater supply of cotton may be sold at reduced prices, an eventuality which is not likely unless important new uses for cotton are developed. It will enhance the profit-making capacity of the few entrepreneurs, and will limit the workers to a small cash income which must be expended for commodities produced elsewhere. Diversification of agriculture is obviously not so profitable on large-scale commercial farms since the profits to the operator of such enterprises come from sales of commercial crops and the operator has no interest in the subsistence crops produced by the laborers and tenants. The family-sized farm and the part-time farm, on the other hand, discard the economies of large-scale buying and management, and substitute the free services of family labor for the efficiency of mechanical cultivation and harvesting.

Social Inertia

The persistence of the plantation as a socio-economic institution surviving the Civil War and subsequent depressions, ramifying in its influence into all phases of southern rural life, indicates that the plantation has evolved to fit a definite socio-economic condition and that it meets the situation better than anything else which can be produced by the unaided efforts of the groups involved. It also indicates the formation of a deeply ingrained set of customs and relationships which will be difficult to change and impossible to revolutionize suddenly. In fact, a search for parallels in European agrarian reform indicates that nations such as Ireland and Denmark, which have accomplished conspicuous tenant reforms, have required several generations to attain their goal.

SPECIFIC PROGRAMS

Measures for relief, rehabilitation, and the fundamental reconstruction of southern agriculture are matters for Federal-

State cooperation, and they have been strongly stimulated by the appropriation of Federal funds. Since it is apparent that these are long time rather than temporary problems, eventual alleviation of some of the conditions set out in this report will be expedited by a continuation of Federal assistance.

In appraising the effect of actual constructive legislation or organization, it is often difficult to separate the faults or virtues of an objective from the excellencies or flaws of the administrative measures designed to attain that objective. Inasmuch as appraisal of administrative procedures is not a function of this chapter, only the objectives are discussed.

State Legislation

More comprehensive State legislation for the protection of the tenant is possible but has never been vigorously sponsored by southern legislatures. The principal proposals which have been made in this field are:

a. Repeal of the laws now on the books of certain States which make it a misdemeanor to quit a contract while in debt. This provision has, in the past more frequently than in the present, led to the use of criminal procedure to hold tenants and a confusion as to the jurisdiction of the Federal anti-peonage laws.

b. Requirement of a written contract with power to call for an accounting on the performance of the agreement before the courts. The power to demand an accounting now exists, but without an aroused public opinion the tenant and the landlord are not equal before the law.

c. Contracts providing the reimbursement of the tenant for permanent improvements of the land or buildings made with his labor and not exhausted before the time of his moving from the holding. Such contracts are necessary if full benefits of programs of soil conservation and permanent improvement of property are to be reaped.

Submarginal Land Retirement Programs

The shifting of population from submarginal lands to more productive lands affects both owner occupants and tenant occupants. However, the owner occupant receives the purchase price of the property and can use his equity to secure a new farm. Unless the tenant is aided toward resettlement in a new location, he is not helped by the procedure.

From the viewpoint of cash crop production control the retirement of submarginal lands in favor of more fertile acres is a two-edged sword. If the more productive acreages are used for cash crops in the same proportion as were the submarginal lands, this will lead to a larger production per acre and to an extent offset the effects of acreage reduction. There is the added dilemma of selecting the method for retiring submarginal lands and determining the most desirable use for the acreage retired from agriculture.

Three methods of promoting the retirement of submarginal lands have been proposed:

a. Purchase by the Federal government and transfer from agricultural use to other uses, such as forestry, public grazing, game preserves, and recreation. The acquisition of all lands which have been judged submarginal would, however, prove prohibitive in cost and would again build up a vast public domain.

b. A legal zoning process in rural areas which would operate similarly to restrictive zoning in cities. This is a process which would have to be carried out State by State and county by county and would encounter many legislative and constitutional snags.

c. A zoning process without legal sanctions which would designate lands unfit for commercial agriculture and by a process of education guide settlers away from these and toward other areas. Such a movement would be supplemented by such measures as the withdrawal of public services from the proscribed areas, the curtailment of road extension and repairs, the abandonment of schools, and encouragement for movement to other sections. This process would be subject to the uneven progress characteristic of programs dependent upon local initiative.

Soil Conservation

More fundamental in the long run are the programs of soil conservation and soil improvement which will restore the productivity of lands now near the margin of profitable use in agriculture. Farmers whose land now provides a precarious existence can be encouraged to put their soil in such condition that it will be more productive. The tenant system as it exists in the South, however, is inimical to soil conservation policies. The tenant system is built around the cotton-corn economy and both of these crops are clean-cultivated row crops which allow the maximum amount of erosion.

Cover crops are usually not cash crops and their profitable use is dependent upon feeding to livestock. Livestock production is difficult, but not impossible, to adapt to a share tenant system.

Diversification of Crops

A program of crop diversification has long been urged by the press and farmers' organizations as a panacea for the ills of the South. On this point again the findings of this study are definitive. The controlling factor on plantations is that the landlord's income is accumulated from his shares of the incomes of a number of tenants and while it may be of advantage to the tenants to produce crops for home use, it is not of so great advantage to the landlord as the production of a cash crop which is readily marketable.

Production Control

The financial benefits of the Agricultural Adjustment Administration program were discussed in chapter VI. The income analysis, however, makes evident the fact that the benefits of price increase, attributed to a number of factors of which production control was one, far outweighed the benefit from acreage reduction payments. These payments averaged $8.00 per cropper as against an increase of $30.00 per bale in the price of each bale of cotton produced (from 6 cents per pound in 1932 to 12 cents in 1935-1936).

The indirect benefits of the adjustment program, if followed through, might be of more permanent value than the direct financial benefits. These indirect benefits come from the pressure toward diversification under the money crop reduction program. With roughly 40 percent of the money crop acreage retired under the A.A.A. program there was a considerable urge to use this land for other purposes. On many farms, land retired from cotton merely lay idle or was sketchily cultivated in food and forage crops. However, idle land under the climatic conditions of the South quickly produces grass, and without excessive expenditure becomes excellent pasture. The increases in pasture shown in the 1935 Census over 1930 indicate this trend. This, coupled with the acquisition of thousands of cattle from the drought area of 1934, gave strong impetus to the development of animal husbandry in the Southeast.

Following the abandonment of the A.A.A. program, the inauguration of soil conservation benefits gave positive stimulus to the use of idle acres for food and forage crops. Under this program, the payment of benefits on food crops produced by the tenant should prove especially effective in raising the standard of living of the tenants. One of the most significant findings of this study is the relationship between increase in tenant income and increase in home use production. Furthermore, this is stable income, not subject to price fluctuations or marketing difficulties. The rapidly increasing farm population of the South emphasizes the difficulty of controlled production in an area of increasing labor supply. Production policies, if they are not to depress the existing standard of living, must be adjusted to an expanding population. Thus soil conservation, diversification, and production control are inextricably interwoven and their success is mutually dependent. All these are hampered by the present tenant system with its objective of producing cash income.

Credit Reform

Federal Land Banks, Cooperative Credit Associations, and feed and seed loans have all exerted an influence toward improvement of the existing credit system by aiding the borrower who had ample security. They have, however, reached the tenant only indirectly since his security is not of the commercial type except where he owns animals subject to a chattel mortgage. The type of credit described later in the section on rehabilitation is more adapted to tenant conditions.

Again, agricultural reconstruction, as embodied in the programs of diversification, soil conservation, and crop control, is impeded by a credit system based on a cash crop as security.

Direct Relief

Relief payments were made to as many as half a million families in the Eastern Cotton States at one time or another and the amount disbursed totalled millions of dollars. These payments alleviated suffering on a broad scale. Such a program was necessary as a palliative for distress but it contributed little to remedy underlying ills. It is significant that work relief was developed and production goods distributed early in the program.

Work Relief

The field work on this study had been completed before the inauguration of the Works Program. Therefore, the direct impact of this program on the plantation was not the object of first-hand observation. Certain facts can, however, be pointed out:

a. A work program is not well adapted to conserving agricultural assets unless it is concentrated in off-seasons, or unless members of farm families other than the operators are available for employment. However, because of effort to operate on submarginal lands, because of large families, or because of natural disaster such as drought or flood, many farm families need cash when the loan of such cash would be economically unsound. Every effort should be made to provide this cash through grants and allow the farmer to remain on his farm and to preserve his agricultural assets.

b. This points to the consideration that in many instances direct relief, such as the Resettlement grants, is most suited to the needs of the farmer. Though perhaps less calculated to preserve his self respect, such grants, nevertheless, leave him free to devote his full time to recouping his farm assets.

c. Work projects which tend to draw farmers into towns and villages should be minimized.

Rural Rehabilitation

The substitution of loans for the purchase of animals and implements, coupled with advice on farm procedures, for direct relief payments marked a distinct advance in the readjustment of dislocated farm families to the land. This program, in effect, advanced the former displaced cropper to the renter status inasmuch as animals and implements were supplied and a rental in kind collected. The first real advance of the tenant above the labor-cropper status comes when he acquires his own animals and implements and consequently receives a larger share of the annual production.

The rehabilitation program also constitutes a major attack on the vicious credit system which is a cornerstone of the old tenant-landlord-merchant system of the Cotton Belt. Extension of easier credit was possible under the rehabilitation program because (1) the risks were spread over thousands of clients, (2) a farm plan was worked out to fit the family composition and land available, and (3) supervision by trained personnel was provided.

Given a period of years, such a program has a fair chance of advancing many tenants to the status of renters of family-sized farms and thus aiding them to take the first step toward ownership. Inasmuch as the program is practically self-liquidating, the money can be used, as the loans are repaid, to absorb other families into the program and extend its benefits.

Promotion of Landownership

The constructive measures discussed up to this point relate to efforts to improve conditions within, or to modify slightly, the existing tenant system.

Landownership promotion proposals involve three other basic principles:

a. Making available to the tenant small family-sized tracts of *good* land. Usually the best commercial crop land is concentrated in the larger holdings which, when sold, are kept in as large tracts as possible and not cut up into family-sized farms. New land brought into cultivation through clearing and stumping, irrigation or drainage, also usually must be developed in large tracts for economy, and it is beyond the means of the small farmer to carry on such operations unaided.

b. Provision of long-time credit on easy terms. The usual period of 3 or 6 years for repayment of mortgages is too short a time for the prospective purchaser to acquire full equity, especially under the unstable conditions faced by the cotton farmer.

   The small cash incomes produced on family-sized cotton farming units emphasize the need of keeping initial costs of these tracts low. Even a 40-year amortization of a $4,000 farm would require payments of $100 per year which would constitute a heavy drain on a cash income such as the 1934 tenant average of about $200.

c. Provision of supervision in the nature of adult education which will not only give the farmer the benefit of improved agricultural practices but will train him in the habits necessary for successful management of his own enterprise.

A program of this kind proposed by the Bankhead-Jones Bill[4] is particularly adapted to the long-time reform of agrarian life in regions of heavy tenant ratios.

The land buying program has been treated separately from the other measures discussed in this section, not because there is conflict between them but because it involves a separate set of mechanisms and objectives. The problems of tenancy in the United States are so far-reaching in significance, and so large in volume, that measures to improve the condition of those remaining within the system should proceed simultaneously with efforts to help the most able tenants to escape from the system.

With a hundred million dollars wholly devoted to supplying $4,000 units, only 25,000 families could be supplied. This is an insignificant proportion of the total tenants in the country. It is evident, therefore, that the contention in the beginning of this section, that the process will of necessity be slow, should be repeated. Also, programs of protection and rehabil-

---

[4] *U. S. 74th Congress, 1st Session, House,* "An Act to Create the Farmers' Home Corporation, etc.," S. 2367, June 26, 1935.

itation should proceed in conjunction with programs of ownership promotion. To promote ownership without alleviating the condition of the mass of tenants would mean working with the upper class only. To endeavor merely to improve and palliate the present tenant system would be like patching a worn-out leaky vessel. The task is sufficiently important, sufficiently challenging, and sufficiently vast to call for a coordinated attack on all fronts.

APPENDIX A

SUPPLEMENTARY TABLES

Table 1—TENANTS ON ALL FARMS AND ON PLANTATIONS IN THE SELECTED PLANTATION AREA OF SEVEN SOUTHEASTERN COTTON STATES, 1910

State	Number of Counties Included	Total Farms			Plantations			Percent Plantation Tenants Were of All Tenants
		Number of Farms	Tenant Operated Farms	Percent Tenancy	Number of Plant-ations	Tenant Farms in Plantations	Number of Tenants per Plantation	
Seven cotton States	270	946,693	653,607	69.0	33,908	355,186	10.5	54.3
Alabama	47	205,969	135,669	65.9	7,287	76,746	10.5	56.6
Arkansas	23	77,749	55,027	70.8	2,674	35,179	13.2	63.9
Georgia	70	161,650	122,488	75.8	6,627	57,003	8.6	46.5
Louisiana	29	73,207	47,823	65.3	2,480	29,654	12.0	62.0
Mississippi	45	200,673	148,785	74.1	7,960	99,432	12.5	66.8
North Carolina	21	79,609	43,729	54.9	1,775	13,548	7.6	31.0
South Carolina	35	147,836	100,086	67.7	5,105	43,624	8.5	43.6

Note: *Census Definition of Tenant Plantation*: "A continuous tract of land of considerable area under the general supervision, or control, of a single individual or firm, all or a part of such tract being divided into at least 5 smaller tracts, which are leased to tenants."

Source: *Thirteenth Census of the United States: 1910*, Vols. VI and VII, and Vol. V, Ch. XII, "Plantations in the South."

Table 2—CHILDREN UNDER 5 YEARS OF AGE PER 1,000 NATIVE-BORN WOMEN 15-44 YEARS OF AGE, IN THE RURAL FARM POPULATION, BY COLOR, IN UNITED STATES AND SEVEN SOUTHEASTERN COTTON STATES, 1920 AND 1930

State	All Races		Native White	Negro
	1920	1930	1930	1930
United States	600	541	529	566
Seven cotton States	658	591	609	568
Alabama	664	610	632	574
Arkansas	621	574	615	488
Georgia	672	578	590	561
Louisiana	632	587	603	568
Mississippi	578	564	600	538
North Carolina	730	632	620	658
South Carolina	686	585	582	589

Source: *Fifteenth Census of the United States: 1930.*

Table 3—TYPE OF TENANTS ON PLANTATIONS, BY AREAS, 1934
(Cotton Plantation Enumeration)

Area	Total	Number of Plantations Operated by				
		Wage Hands	Croppers	Other Share Tenants	Renters	Mixed
Total: Number	646	27	100	20	37	462
Total: Percent	100.0	4.2	15.5	3.1	5.7	71.5
Atlantic Coast Plain	56	6	12	-	-	38
Upper Piedmont	40	1	3	1	-	35
Black Belt[a]	112	2	15	3	5	87
Black Belt[b]	99	5	4	1	17	72
Upper Delta	133	5	33	8	-	87
Lower Delta	50	-	7	3	12	28
Muscle Shoals	22	1	-	2	1	18
Interior Plain	30	-	8	1	-	21
Mississippi Bluffs	47	1	11	1	2	32
Red River	28	5	5	-	-	18
Arkansas River	29	1	2	-	-	26

[a] Cropper and other share tenant majority.
[b] Renter majority.

Table 4—COLOR OF TENANTS ON PLANTATIONS, BY AREAS, 1934
(Cotton Plantation Enumeration)

Area	Number of Plantations by Color of Tenants			
	Total	White	Negro	Both White and Negro
Total: Number	646	31	341	274
Total: Percent	100.0	4.8	52.8	42.4
Atlantic Coast Plain	56	1	26	29
Upper Piedmont	40	10	4	26
Black Belt (A)[a]	112	3	54	55
Black Belt (B)[b]	99	3	62	34
Upper Delta	133	3	91	39
Lower Delta	50	–	37	13
Muscle Shoals	22	7	6	9
Interior Plain	30	–	15	15
Mississippi Bluffs	47	3	26	18
Red River	28	1	11	16
Arkansas River	29	–	9	20

[a]Cropper and other share tenant majority.
[b]Renter majority.

Table 5—LAND PROPRIETORSHIPS, BY SIZE, IN SELECTED COUNTIES OF NORTH CAROLINA,[a] GEORGIA,[b] AND MISSISSIPPI,[c] 1934

Size of Proprietorship	Number	Percent
Total	79,456	100.0
3 to 10 acres	4,058	5.1
10 to 20 acres	4,949	6.2
20 to 50 acres	18,767	23.6
50 to 100 acres	20,240	25.5
100 to 175 acres	15,242	19.2
175 to 260 acres	6,590	8.3
260 to 500 acres	6,122	7.7
500 acres and over	3,488	4.4

[a]Cleveland, Edgecombe, Greene, Halifax, Iredell, Johnston, Lincoln, Robeson, Rutherford, Union, and Wilson.
[b]Banks, Berrien, Burke, Butts, Clay, Coweta, Decatur, Dougherty, Forsyth, Greene, Jasper, Johnson, Lincoln, Lowndes, Madison, Newton, Paulding, Polk, Putnam, Sumter, Talbot, Telfair, Troup, and Wilkes.
[c]Adams, Carroll, Clay, Coahoma, Quitman, Warren, Washington, and Yazoo.
Source: Tax digests in the respective counties.

Table 6—LAND PROPRIETORSHIPS, BY SIZE AND BY AREAS,[a] 1911, 1922, AND 1934

Area	Year[b]	Tracts by Acreage								
		Total	3–10 Acres	10–20 Acres	20–50 Acres	50–100 Acres	100–175 Acres	175–260 Acres	260–500 Acres	500 Acres and Over
						Number				
Black Belt	1922	15,840	541	586	2,250	3,817	3,789	1,960	1,657	1,240
	1934	19,675	1,452	1,191	3,379	4,718	4,056	2,034	1,682	1,163
Upper Piedmont	1922	15,930	292	532	3,380	5,151	3,696	1,419	1,034	426
	1934	18,482	604	996	4,512	5,759	3,766	1,464	937	444
Delta	1911	9,839	415	483	2,363	2,052	1,859	808	1,156	703
	1934	10,406	480	671	2,599	2,249	1,804	768	1,203	632
Mississippi Bluffs	1911	8,563	307	322	1,811	2,364	1,852	777	811	319
	1934	8,592	330	359	2,640	2,079	1,644	652	693	195
Atlantic Coast Plain	1922	9,761	253	589	2,334	2,738	1,991	800	687	369
	1934	13,050	547	1,153	3,433	3,444	2,220	995	798	460
						Percent				
Black Belt	1922	100.0	3.4	3.7	14.2	24.1	23.9	12.4	10.5	7.8
	1934	100.0	7.4	6.1	17.2	24.0	20.6	10.3	8.5	5.9
Upper Piedmont	1922	100.0	1.8	3.3	21.2	32.4	23.2	8.9	6.5	2.7
	1934	100.0	3.3	5.4	24.4	31.2	20.4	7.9	5.0	2.4
Delta	1911	100.0	4.2	4.9	24.0	21.0	18.9	8.2	11.7	7.1
	1934	100.0	4.6	6.4	25.0	21.6	17.3	7.4	11.6	6.1
Mississippi Bluffs	1911	100.0	3.6	3.8	21.1	27.6	21.6	9.1	9.5	3.7
	1934	100.0	3.8	4.2	30.7	24.2	19.1	7.6	8.1	2.3
Atlantic Coast Plain	1922	100.0	2.6	6.0	23.9	28.1	20.4	8.2	7.0	3.8
	1934	100.0	4.2	8.8	26.4	26.4	17.0	7.6	6.1	3.5

[a]Sample counties included: — Black Belt: North Carolina - Union and Robeson; Georgia - Butts, Clay, Greene, Jasper, Johnson, Lincoln, Putnam, Sumter, Talbot, Troup, Wilkes, and Dougherty; Mississippi - Clay. Upper Piedmont: North Carolina - Rutherford, Cleveland, Lincoln, and Iredell; Georgia - Banks, Coweta, Forsyth, Madison, Newton, Paulding, and Polk. Delta: Mississippi - Adams, Coahoma, Quitman, Warren, and Washington. Mississippi Bluffs: Mississippi - Carroll and Yazoo. Atlantic Coast Plain: North Carolina - Edgecombe, Wilson, Johnston, Greene, and Halifax; Georgia - Telfair.

[b]Owing to the unsatisfactory nature of the data for 1922, the year 1911 was used as the base for the Delta and Mississippi Bluffs areas.

Source: Tax digests in the respective counties.

Table 7—OPERATORS OWNING OTHER FARMS AND NUMBER OF OTHER FARMS OWNED, BY AREAS, 1934
(Cotton Plantation Enumeration)

Area	Total Operators	Operators Owning Other Farms	Total Number of Other Farms Owned	Average Number of Other Farms per Operator
Total	646	251	723	2.9
Atlantic Coast Plain	56	32	64	2.0
Upper Piedmont	40	20	52	2.6
Black Belt (A)[a]	112	50	179	3.6
Black Belt (B)[b]	99	49	123	2.5
Upper Delta	133	31	133	4.3
Lower Delta	50	21	39	1.9
Muscle Shoals	22	4	7	1.8
Interior Plain	30	11	35	3.2
Mississippi Bluffs	47	12	27	2.3
Red River	28	10		1.9
Arkansas River	29	11	5	4.1

[a]Cropper and other share tenant majority.

[b]Renter majority.

Table 8—LAND IN FARMS HELD BY CORPORATIONS IN SELECTED COUNTIES OF NORTH CAROLINA, GEORGIA, AND MISSISSIPPI, 1934

Item	North Carolina[a]		Georgia[b]		Mississippi[c]	
	Acres of Farm Land		Acres of Farm Land		Acres of Farm Land	
	Number	Percent	Number	Percent	Number	Percent
Total farm land held by corporations	434,428	100.0	475,189	100.0	181,979	100.0
Land banks	87,072	20.0	74,634	15.7	96,664	53.2
Depository banks	60,026	13.8	106,456	22.4	18,257	10.0
Insurance companies	48,071	11.1	253,433	53.3	60,249	33.1
All other corporations	239,259	55.1	40,666	8.6	6,809	3.7
Total land in farms, 1935[d]	4,426,990		4,353,018		2,149,860	
Percent of total land in farms held by corporations	9.8		10.9		8.5	

[a] Union, Robeson, Anson, Scotland, Rutherford, Cleveland, Lincoln, Iredell, Catawba, Edgecombe, Wilson, Johnston, Greene, Halifax, Lenoir, and Sampson.

[b] Butts, Clay, Greene, Jasper, Johnson, Lincoln, Putnam, Sumter, Talbot, Troup, Wilkes, Dougherty, Banks, Coweta, Forsyth, Madison, Newton, Paulding, Polk, Telfair, Burke, Decatur, Berrien, and Lowndes.

[c] Coahoma, Quitman, Washington, Carroll, Yazoo, and Clay.

[d] *United States Census of Agriculture: 1935*, Preliminary Report.

Source: Tax digests in the respective counties.

Table 9—ACRES IN PLANTATIONS, BY DATE OF ACQUISITION, BY AREAS, 1934
(Cotton Plantation Enumeration)

Area	Total Plantations Reporting[a]	Number of Acres by Date of Acquisition																	
		Total[b]		Before 1900		1900 - 1905		1905 - 1910		1910 - 1915		1915 - 1920		1920 - 1925		1925 - 1930		1930 - 1935	
		Number	Percent	Number	Percent	Number	Percent	Number	Percent	Number	Percent	Number	Percent	Number	Percent	Number	Percent	Number	Percent
Total	534	450,737	100.0	34,134	7.6	29,172	6.5	30,005	6.7	36,258	8.0	66,225	14.7	69,053	15.3	90,294	20.0	95,596	21.2
Atlantic Coast Plain	45	40,257	100.0	647	1.6	716	1.8	2,335	5.8	2,033	5.0	3,223	8.0	9,010	22.4	14,907	37.1	7,386	18.3
Upper Piedmont	34	12,738	100.0	543	4.3	1,667	13.1	2,014	15.8	1,949	15.3	2,135	16.7	1,409	11.1	1,856	14.6	1,165	9.1
Black Belt (A)[c]	98	73,359	100.0	6,362	8.7	1,283	1.7	3,321	4.5	5,161	7.0	13,459	18.4	7,959	10.9	24,225	33.0	11,589	15.8
Black Belt (B)[d]	86	66,946	100.0	6,232	9.3	4,287	6.4	5,194	7.8	8,037	12.0	7,417	11.1	15,297	22.8	6,789	10.1	13,693	20.5
Upper Delta	101	93,104	100.0	2,717	2.9	2,295	2.5	5,032	5.4	6,539	7.0	9,758	10.5	23,928	25.7	11,371	12.2	31,464	33.8
Lower Delta	44	41,462	100.0	9,516	22.9	8,030	19.4	2,803	6.8	5,797	14.0	4,799	11.6	4,344	10.5	2,468	5.9	3,705	8.9
Muscle Shoals	17	8,369	100.0	980	11.7	160	1.9	1,485	17.7	585	7.0	2,000	23.9	1,009	12.1	1,135	13.6	1,015	12.1
Interior Plain	20	26,007	100.0	2,578	9.9	812	3.1	2,005	7.7	317	1.2	5,682	21.9	370	1.4	7,030	27.1	7,213	27.7
Mississippi Bluffs	44	30,520	100.0	1,921	6.3	5,575	18.3	276	0.9	2,194	7.2	8,339	27.3	1,663	5.4	2,561	8.4	7,991	26.2
Red River	19	13,628	100.0	417	3.1	975	7.1	-	-	640	4.7	4,600	33.8	2,669	19.6	3,223	23.6	1,104	8.1
Arkansas River	26	44,347	100.0	2,221	5.0	3,372	7.6	5,540	12.5	3,006	6.8	4,813	10.9	1,395	3.1	14,729	33.2	9,271	20.9

[a] Data not available for 112 plantations.
[b] Land in crops, tillable land idle, pasture land, woods not pastured, and waste land.
[c] Cropper and other share tenant majority.
[d] Renter majority.

Table 10—RESIDENT, SEMI-ABSENTEE, AND ABSENTEE OPERATORS, BY AREAS, 1934
(Cotton Plantation Enumeration)

Area	Number of Plantations by Type of Operator			
	Total	Resident[a]	Semi-absentee[b]	Absentee[c]
Total: Number	646	551	55	40
Total: Percent	100.0	85.3	8.5	6.2
Atlantic Coast Plain	56	45	8	3
Upper Piedmont	40	32	6	2
Black Belt (A)[d]	112	95	8	9
Black Belt (B)[e]	99	77	9	13
Upper Delta	133	123	5	5
Lower Delta	50	38	8	4
Muscle Shoals	22	20	1	1
Interior Plain	30	28	2	0
Mississippi Bluffs	47	39	7	1
Red River	28	26	1	1
Arkansas River	29	28	0	1

[a]Operator was classified as resident if he lived on the plantation, visited it daily, or hired an overseer.

[b]Operator was classified as semi-absentee if he lived within 10 miles of the plantation and visited it as often as once a week.

[c]Operator was classified as absentee if he lived more than 10 miles from the plantation, did not hire an overseer, and visited the plantation less frequently than once a week.

[d]Cropper and other share tenant majority.

[e]Renter majority.

Table 11—VALUE OF LAND, BUILDINGS, ANIMALS, AND MACHINERY, BY AREAS, 1934
(Cotton Plantation Enumeration)

Area	Total Plantations Reporting[a]	Value					Value per Plantation				
		Total	Land	Buildings[b]	Animals	Machinery	Total	Land	Buildings	Animals	Machinery
Total	632	$18,134,300	$13,700,900	$2,503,400	$1,181,200	$748,800	$28,694	$21,679	$ 3,961	$1,869	$1,185
Atlantic Coast Plain	56	1,040,800	732,400	190,600	84,900	32,900	18,586	13,079	3,403	1,516	588
Upper Piedmont	39	514,300	365,100	83,700	44,500	21,000	13,187	9,362	2,146	1,141	538
Black Belt (A)[c]	111	1,985,400	1,461,500	283,900	156,800	83,200	17,886	13,166	2,558	1,413	749
Black Belt (B)[d]	93	1,372,600	1,002,200	185,100	149,400	35,900	14,759	10,776	1,990	1,607	386
Upper Delta	132	5,642,400	4,114,600	916,700	313,700	297,400	42,745	31,171	6,945	2,376	2,253
Lower Delta	49	873,700	660,800	107,900	69,800	35,200	17,831	13,486	2,202	1,425	718
Muscle Shoals	22	398,500	334,300	36,900	19,800	7,500	18,114	15,196	1,677	900	341
Interior Plain	30	1,395,500	1,150,800	146,900	62,100	35,700	46,517	38,360	4,897	2,070	1,190
Mississippi Bluffs	46	1,003,800	785,500	108,100	73,800	36,400	21,822	17,076	2,350	1,605	791
Red River	28	1,697,000	1,361,900	171,000	94,100	70,000	60,607	48,639	6,107	3,361	2,500
Arkansas River	26	2,210,300	1,731,800	272,600	112,300	93,600	85,012	66,608	10,485	4,319	3,600

[a]Data not available for 14 plantations.

[b]Value of operator's residence, gins, and commissaries excluded. Enumerators were instructed to "enter values at conservative market value, not low assessed value or high speculative value."

[c]Cropper and other share tenant majority.

[d]Renter majority.

Table 12—CROP EXPENDITURES ON PLANTATIONS, BY AREAS, 1934
(Cotton Plantation Enumeration)

Area	Total Plantations	Total Expenses	Fertilizer	Feed	Interest	Expenditures per Plantation									
						Amount					Percent				
						Total	Fertilizer	Feed	Interest	All Other	Total	Fertilizer	Feed	Interest	All Other
Total	646	$2,243,006	$217,148	$96,457	$44,500	$ 3,472	$336	$149	$ 69	$2,918	100.0	9.7	4.3	2.0	84.0
Atlantic Coast Plain	56	164,944	54,698	4,667	2,001	2,945	977	83	36	1,849	100.0	33.2	2.8	1.2	62.8
Upper Piedmont	40	51,260	16,473	1,511	746	1,282	412	38	19	813	100.0	32.1	3.0	1.5	63.4
Black Belt (A)[a]	112	195,991	41,682	9,824	4,326	1,750	372	88	39	1,251	100.0	21.3	5.0	2.2	71.5
Black Belt (B)[b]	99	130,816	28,324	5,104	3,974	1,321	286	52	40	943	100.0	21.7	3.9	3.0	71.4
Upper Delta	133	869,614	37,822	34,889	11,560	6,538	284	262	87	5,905	100.0	4.3	4.0	1.3	90.4
Lower Delta	50	50,760	2,282	2,267	1,812	1,015	46	45	36	888	100.0	4.5	4.5	3.5	87.5
Muscle Shoals	22	22,087	1,486	359	476	1,004	68	16	22	898	100.0	6.8	1.6	2.2	89.4
Interior Plain	30	97,299	5,094	7,319	1,036	3,243	170	244	35	2,794	100.0	5.2	7.5	1.1	86.2
Mississippi Bluffs	47	153,292	18,688	3,068	7,370	3,262	398	65	157	2,642	100.0	12.2	2.0	4.8	81.0
Red River	28	201,534	3,325	9,940	5,172	7,196	119	355	185	6,539	100.0	1.7	4.9	2.6	90.8
Arkansas River	29	305,409	7,274	17,509	6,027	10,531	251	604	208	9,468	100.0	2.4	5.7	2.0	89.9

[a] Cropper and other share tenant majority.
[b] Renter majority.

Table 13—PLANTATIONS WITH AND WITHOUT COMMISSARIES, BY AREAS, 1934
(Cotton Plantation Enumeration)

Area	Total Plantations Reporting[a]		Without Commissary		With Compulsory Commissary		Without Conpulsory Commissary	
	Number	Percent	Number	Percent	Number	Percent	Number	Percent
Total	634	100.0	470	74.1	95	15.0	69	10.9
Atlantic Coast Plain	56	100.0	46	82.1	9	16.1	1	1.8
Upper Piedmont	40	100.0	36	90.0	3	7.5	1	2.5
Black Belt (A)[b]	112	100.0	97	86.6	10	8.9	5	4.5
Black Belt (B)[c]	99	100.0	80	80.8	13	13.1	6	6.1
Upper Delta	121	100.0	78	64.5	16	13.2	27	22.3
Lower Delta	50	100.0	38	76.0	3	6.0	9	18.0
Muscle Shoals	22	100.0	21	95.5	1	4.5	-	-
Interior Plain	30	100.0	20	66.7	4	13.3	6	20.0
Mississippi Bluffs	47	100.0	34	72.4	8	17.0	5	10.6
Red River	28	100.0	12	42.9	10	35.7	6	21.4
Arkansas River	29	100.0	8	27.6	18	62.1	3	10.3

[a]Data not available for 32 plantations.
[b]Cropper and other share tenant majority.
[c]Renter majority.

Table 14-A—SOCIAL CONTRIBUTIONS OF PLANTATIONS, BY AREAS, 1934
(Cotton Plantation Enumeration)

Area	Total Plantations	Plantations Reporting Contributions	Percent of Total	Plantations Reporting Actual Amount of Contribution[a]	Amount of Contributions					Average per Plantation Reporting Amount of Contribution				
					Total	Doctor	School	Church	Entertainment	Total	Doctor	School	Church	Entertainment
Total	646	199	30.8	144	$8,705	$5,729	$259	$1,823	$894	$ 60.45	$ 39.78	$1.80	$12.66	$ 6.21
Atlantic Coast Plain	56	8	14.3	6	105	95	-	-	10	17.50	15.83	-	-	1.67
Upper Piedmont	40	9	22.5	8	320	120	10	105	85	40.00	15.00	1.25	13.13	10.62
Black Belt (A)[b]	112	30	26.8	27	1,467	1,228	42	60	137	54.33	45.48	1.56	2.22	5.07
Black Belt (B)[c]	99	41	41.4	40	1,565	932	22	611	-	39.13	23.30	0.55	15.28	-
Upper Delta	133	48	36.1	20	2,100	1,565	110	155	270	105.00	78.25	5.50	7.75	13.50
Lower Delta	50	8	16.0	6	190	130	25	35	-	31.67	21.67	4.17	5.83	-
Muscle Shoals	22	3	13.6	3	315	315	-	-	-	105.00	105.00	-	-	-
Interior Plain	30	9	30.0	6	575	300	25	210	40	95.83	50.00	4.17	35.00	6.66
Mississippi Bluffs	47	19	40.4	9	667	448	-	82	137	74.11	49.78	-	9.11	15.22
Red River	28	7	25.0	5	595	245	-	350	-	119.00	49.00	-	70.00	-
Arkansas River	29	17	58.6	14	806	351	25	215	215	57.57	25.07	1.78	15.36	15.36

[a]For 55 of the 199 plantations reporting a contribution the actual amount of the contribution was not available.
[b]Cropper and other share tenant majority.
[c]Renter majority.

Table 14-B—SOCIAL CONTRIBUTIONS OF PLANTATIONS, BY AREAS, 1934
(Cotton Plantation Enumeration)

Area	Total Plantations	Number of Plantations				
		Paying Fines	Serving as Parole Sponsor	Allowing Use of Plantation Owned Transportation		
				Total	Once a Month or Oftener	Less than Once a Month
Total	646	137	72	489	349	140
Atlantic Coast Plain	56	7	5	43	42	1
Upper Piedmont	40	11	6	33	26	7
Black Belt (A)[a]	112	14	8	95	76	19
Black Belt (B)[b]	99	14	10	83	66	17
Upper Delta	133	37	16	97	60	37
Lower Delta	50	7	6	20	11	9
Muscle Shoals	22	4	-	17	16	1
Interior Plain	30	19	3	28	8	20
Mississippi Bluffs	47	6	3	30	23	7
Red River	28	11	4	24	9	15
Arkansas River	29	7	11	19	12	7

[a]Cropper and other share tenant majority.
[b]Renter majority.

Table 15—USE OF LAND IN PLANTATIONS, BY AREAS, 1934
(Cotton Plantation Enumeration)

Area	Total Plantations	Total Acreage		Crops		Idle		Pasture		Woods		Waste	
		Number	Percent	Number	Percent	Number	Percent	Number	Percent	Number	Percent	Number	Percent
Total	646	586,042	100.0	248,513	42.4	41,009	7.0	104,392	17.8	138,509	23.6	53,619	9.2
Atlantic Coast Plain	56	43,979	100.0	16,473	37.5	4,459	10.1	3,703	8.4	13,947	31.7	5,397	12.3
Upper Piedmont	40	17,474	100.0	8,459	48.4	1,140	6.5	3,209	18.4	3,559	20.4	1,107	6.3
Black Belt (A)[a]	112	87,904	100.0	30,812	35.0	6,937	7.9	17,593	20.0	27,582	31.4	4,980	5.7
Black Belt (B)[b]	99	83,161	100.0	25,361	30.5	6,628	8.0	31,105	37.4	13,353	16.0	6,714	8.1
Upper Delta	133	137,111	100.0	74,873	54.6	7,536	5.5	11,751	8.6	32,240	23.5	10,711	7.8
Lower Delta	50	57,319	100.0	10,337	18.0	4,671	8.1	17,810	31.1	16,911	29.5	7,590	13.3
Muscle Shoals	22	12,200	100.0	4,943	40.5	2,691	22.0	1,730	14.2	1,913	15.7	923	7.6
Interior Plain	30	34,797	100.0	15,680	45.1	2,115	6.1	4,560	13.1	8,918	25.6	3,524	10.1
Mississippi Bluffs	47	36,930	100.0	17,769	48.1	1,740	4.7	4,219	11.4	8,847	24.0	4,355	11.8
Red River	28	25,229	100.0	14,865	58.9	1,044	4.1	4,391	17.4	2,663	10.6	2,266	9.0
Arkansas River	29	49,938	100.0	28,941	57.9	2,048	4.1	4,321	8.7	8,576	17.2	6,052	12.1

[a] Cropper and other share tenant majority.
[b] Renter majority.

Table 16—CROP ACRES IN PLANTATIONS, BY TENURE STATUS OF RESIDENT FAMILIES, BY AREAS, 1934
(Cotton Plantation Enumeration)

Area	Total Plantations Reporting[a]	Number of Crop Acres					Percent of Total Crop Acres					Number of Families					Crop Acres per Family				
		Total	Wage Hands	Croppers	Other Share Tenants	Renters	Total	Wage Hands	Croppers	Other Share Tenants	Renters	Total	Wage Hands	Croppers	Other Share Tenants	Renters	Total	Wage Hands	Croppers	Other Share Tenants	Renters
Total	442	224,235	67,203	102,565	34,898	19,569	100.0	30.0	45.7	15.6	8.7	8,886	1,501	5,234	1,369	782	25	45	20	25	25
Atlantic Coast Plain	44	15,415	6,861	6,590	849	1,115	100.0	44.5	42.8	5.5	7.2	441	186	227	16	12	35	37	29	53	93
Upper Piedmont	25	7,223	2,151	3,159	1,804	109	100.0	29.8	43.7	25.0	1.5	240	57	120	54	9	30	38	26	33	12
Black Belt (A)[b]	74	25,959	9,076	12,082	2,098	2,703	100.0	35.0	46.5	8.1	10.4	839	189	448	91	111	31	48	27	23	24
Black Belt (B)[c]	74	22,556	7,183	6,581	727	8,065	100.0	31.8	29.2	3.2	35.8	817	232	239	29	317	28	31	28	25	25
Upper Delta	77	65,905	16,502	34,806	15,087	1,510	100.0	25.0	52.8	19.9	2.3	3,013	326	2,013	634	40	22	51	17	21	38
Lower Delta	34	8,788	433	3,685	1,539	3,131	100.0	4.9	42.0	17.5	35.6	545	19	249	86	191	16	23	15	18	16
Muscle Shoals	17	4,651	894	653	1,794	1,310	100.0	19.2	14.0	38.6	28.2	139	24	25	55	35	33	37	26	33	37
Interior Plain	18	15,478	2,475	7,962	5,041	-	100.0	16.0	51.4	32.6	-	408	32	256	120	-	38	77	31	42	-
Mississippi Bluffs	36	17,134	3,085	9,374	3,091	1,584	100.0	18.0	54.7	18.1	9.2	723	69	464	125	65	24	45	20	25	24
Red River	17	12,743	5,678	5,262	1,773	30	100.0	44.6	41.3	13.9	0.2	547	189	300	57	1	23	30	18	31	30
Arkansas River	26	28,383	12,865	12,411	3,095	12	100.0	45.3	43.7	10.9	0.1	1,174	178	893	102	1	24	72	14	30	12

[a]Data not available for 208 plantations.
[b]Cropper and other share tenant majority.
[c]Renter majority.

Table 17—RATIO OF PRICES RECEIVED FOR COTTON AND COTTON-SEED AND FOR ALL AGRICULTURAL COMMODITIES TO PRICES PAID FOR COMMODITIES BOUGHT, 1924-1935

Year	For Cotton and Cotton-seed	For All Agricultural Commodities
1924	139	94
1925	113	99
1926	79	94
1927	84	91
1928	98	96
1929	94	95
1930	70	87
1931	51	70
1932	44	61
1933	59	64
1934	80	73
1935	81[a]	86[a]

[a]Preliminary.

Source: *The Agricultural Situation, January 1, 1936*, U. S. Bureau of Agricultural Economics, Vol. 20, No. 1, p. 21.

Table 18—AVERAGE NET DEMAND AND TIME DEPOSITS IN BANKS LOCATED IN TOWNS OF LESS THAN 15,000 POPULATION IN UNITED STATES AND SEVEN SOUTHEASTERN COTTON STATES, 1929-1935

Year	United States	Seven Cotton States
1929	100	100
1930	94	91
1931	80	49
1932	65	37
1933	57	31
1934	52	38
1935	62	53

Source: *Annual Reports of the Federal Reserve Board*, as of January 1.

Table 19—PLANTATION LIVESTOCK, BY AREAS, 1934
(Cotton Plantation Enumeration)

Area	Total Plantations	Total Livestock					Number per Plantation				
		Mules and Horses	Cows	Calves	Pigs	Chickens	Mules and Horses	Cows	Calves	Pigs	Chickens
Total	646	8.829	11,666[c]	4,517[c]	8,733	31,961	14	18	7	14	49
Atlantic Coast Plain	56	547	528	285	959	3,305	10	9	5	17	59
Upper Piedmont	40	274	228	128	163	1,862	7	6	3	4	47
Black Belt (A)[a]	112	981	2,331[d]	1,072[d]	1,757	5,325	9	21	10	16	48
Black Belt (B)[b]	99	764	3,556	1,300	1,024	4,846	8	36	13	10	49
Upper Delta	133	2,604	1,732[e]	238[e]	2,051	6,280	20	13	2	15	47
Lower Delta	50	414	1,279	391	669	2,687	8	26	8	13	54
Muscle Shoals	22	109	117	135	99	1,100	5	5	6	5	50
Interior Plain	30	486	309[f]	93[f]	460	1,865	16	10	3	15	62
Mississippi Bluffs	47	740	685	333	559	2,532	16	15	7	12	54
Red River	28	746	621[g]	435[g]	424	1,205	27	22	16	15	43
Arkansas River	29	1,164	280[h]	107[h]	568	954	40	10	4	20	33

[a] Cropper and other share tenant majority.
[b] Renter majority.
[c] 1,456 "cows and calves" not shown.
[d] 87 "cows and calves" not shown.
[e] 86 "cows and calves" not shown.
[f] 527 "cows and calves" not shown.
[g] 736 "cows and calves" not shown.
[h] 20 "cows and calves" not shown.

Table 20—OPERATORS' LONG TERM DEBTS, BY AREAS, 1934
(Cotton Plantation Enumeration)

Area	Total Oper-ators	Operators Reporting Debt		Total Debts	Debt per Operator Report-ing Debt	Percent of Operators Reporting Debt					
		Num-ber	Per-cent			Mort-gage	Bank	Mer-chant Note	Open Account	Govern-ment	Other
Total	646	284	44	$3,330,160	$11,726	88.0	6.7	2.5	2.1	1.8	4.2
Atlantic Coast Plain	56	22	39	137,275	6,240	100.0	–	–	–	–	–
Upper Piedmont	40	21	53	104,938	4,997	100.0	–	4.8	–	–	–
Black Belt (A)[a]	112	54	48	380,950	7,055	90.7	1.9	1.9	3.7	–	1.9
Black Belt (B)[b]	99	37	37	154,743	4,182	81.1	5.4	2.7	10.8	–	5.4
Upper Delta	133	48	36	756,989	15,771	93.8	4.2	4.2	–	–	–
Lower Delta	50	20	40	141,543	7,077	70.0	25.0	5.0	–	10.0	–
Muscle Shoals	22	5	23	18,900	3,780	40.0	–	–	–	–	60.0
Interior Plain	30	16	53	134,790	8,424	93.8	–	–	–	6.3	6.3
Mississippi Bluffs	47	22	47	187,410	8,519	95.5	4.5	–	–	–	4.5
Red River	28	19	68	563,133	29,639	68.4	21.1	5.3	–	1.6	1.6
Arkansas River	29	20	69	749,489	37,474	90.0	15.0	–	–	–	5.0

[a] Cropper and other share tenant majority.
[b] Renter majority.

Table 21—FARM MORTGAGE DEBT, BY TENURE OF OPERATORS, IN UNITED STATES[a] AND SEVEN SOUTHEASTERN COTTON STATES, 1910, 1920, 1925, AND 1928

Area and Tenure	Estimated Farm Mortgage Debt (in thousands of dollars)				Percent Increase[c]		
	1910	1920	1925	1928	1910–1920	1920–1925	1925–1928
United States							
Total	$3,320,470	$7,857,700	$9,360,620	$9,468,526	136.6	19.1	1.2
Owners[b]	2,197,800	5,314,150	5,504,437	5,560,017	141.8	3.6	1.0
Tenants	977,730	2,185,480	3,612,193	3,644,009	123.5	65.3	0.9
Managers	144,940	358,070	243,990	264,500	147.0	-31.9	8.4
Seven cotton states							
Total	165,780	442,630	588,092	637,597	167.0	32.9	8.4
Owners[b]	88,920	241,620	308,929	331,363	171.7	27.9	7.3
Tenants	69,700	182,360	257,645	280,494	161.6	41.3	8.9
Managers	7,160	18,650	21,518	25,740	160.5	15.4	19.6

[a] In addition to the farm mortgage debt as reported by the Census, the estimated farm mortgage debt on other farms has been added.
[b] Includes all part owners.
[c] (–) indicates a decrease.

Source: Wickens, David L., *Farm Mortgage Credit*, U. S. Department of Agriculture, Technical Bulletin 288, February 1932, pp. 3–4, and Appendix pp. 96–97.

Table 22—RATIO OF MORTGAGE DEBT TO VALUE OF ALL FARMS IN UNITED STATES AND SEVEN SOUTHEASTERN COTTON STATES, JANUARY 1, 1910, 1920, 1925, AND 1928

State	Ratio of Debt to Value of All Farms (Percent)			
	1910	1920	1925	1928
United States	9.5	11.8	18.9	21.0
Alabama	8.6	10.2	16.0	17.8
Arkansas	7.2	10.2	18.1	20.8
Georgia	6.0	7.4	18.6	23.9
Louisiana	8.0	8.7	17.8	20.3
Mississippi	9.4	9.8	23.6	26.9
North Carolina	4.2	5.3	8.5	10.5
South Carolina	6.2	6.3	15.0	21.2

Source: Wickens, David L., *Farm Mortgage Credit*, U. S. Department of Agriculture Technical Bulletin 288, February 1932, p. 50.

Table 23—FREQUENCY OF FARM MORTGAGE DEBT IN UNITED STATES AND SEVEN SOUTHEASTERN COTTON STATES, BY TENURE OF OPERATOR, JANUARY 1, 1935, AND 1928

State	All Farms		Full Owner Operated Farms		Part Owner Operated Farms		Tenant Operated Farms	
	1925	1928	1925	1928	1925	1928	1925	1928
United States	34.8	36.0	34.0	34.7	48.1	48.5	32.5	34.8
Alabama	28.9	30.5	29.8	30.6	30.2	31.3	28.2	30.3
Arkansas	33.7	36.7	32.5	34.9	35.4	38.0	34.3	37.8
Georgia	29.6	31.5	27.2	26.2	27.2	26.7	30.6	34.3
Louisiana	28.2	30.8	27.0	29.3	32.8	35.6	28.5	31.4
Mississippi	31.9	33.7	33.0	33.9	34.8	36.0	31.2	33.5
North Carolina	20.1	22.5	18.8	20.8	22.2	25.1	21.1	23.9
South Carolina	28.2	31.0	26.1	27.5	24.5	26.3	29.3	32.9

Source: Wickens, David L., *Farm Mortgage Credit*, U.S. Department of Agriculture Technical Bulletin 288, February 1932, p. 43.

Table 24—FARM LAND HELD BY CORPORATIONS IN 46 NORTH CAROLINA,[a] GEORGIA,[b] AND MISSISSIPPI[c] COUNTIES, 1934

Item	Acres	Percent
Total land in farms	10,929,868	
Total land held by corporations	1,091,596	100.0
Land banks	258,370	23.7
Depository banks	184,739	16.9
Insurance companies	361,753	33.1
Other	286,734	26.3

[a]Union, Robeson, Anson, Scotland, Rutherford, Cleveland, Lincoln, Iredell, Catawba, Edgecombe, Wilson, Johnston, Greene, Halifax, Lenoir, and Sampson.

[b]Butts, Clay, Greene, Jasper, Johnson, Lincoln, Putnam, Sumter, Talbot, Troup, Wilkes, Dougherty, Banks, Coweta, Forsyth, Madison, Newton, Paulding, Polk, Telfair, Burke, Decatur, Berrien, and Lowndes.

[c]Coahoma, Quitman, Washington, Carroll, Yazoo, and Clay.

Source: Tax digests in the respective counties.

Table 25—AMOUNT AND ANNUAL RATE OF INTEREST OF GOVERNMENT, MERCHANT, FERTILIZER, AND BANK LOANS, BY AREAS, 1934
(Cotton Plantation Enumeration)

Area	Total Plantations	Government Loans			Merchant Loans			Fertilizer Loans			Bank Loans		
		Plantations Reporting Loans	Total Amount of Loans	Average Annual Rate of Interest	Plantations Reporting Loans	Total Amount of Loans	Average Annual Rate of Interest	Plantations Reporting Loans	Total Amount of Loans	Average Annual Rate of Interest	Plantations Reporting Loans	Total Amount of Loans	Average Annual Rate of Interest
Total	646	57	$164,214	10.4	48[c]	$90,866	16.4	7	$3,422	21.1	225[d]	$484,066	15.2
Atlantic Coast Plain	56	6	5,500	10.5	4	6,200	36.8	1	100	5.4	19	27,250	11.7
Upper Piedmont	40	4	6,775	9.5	3	1,700	26.5	–	–	–	5	2,100	18.7
Black Belt (A)[a]	112	1	300	8.0	11	12,910	22.3	4	1,642	27.2	44	35,725	17.0
Black Belt (B)[b]	99	7	10,600	12.8	9	7,075	16.2	2	1,680	14.0	24	23,250	21.2
Upper Delta	133	11	43,650	11.0	7	14,443	10.5	–	–	–	52	153,136	14.8
Lower Delta	50	2	2,910	14.5	6	7,090	11.0	–	–	–	12	15,400	13.9
Muscle Shoals	22	–	–	–	–	–	–	–	–	–	10	6,325	20.1
Interior Plain	30	9	10,295	6.5	1	2,000	10.0	–	–	–	11	19,605	11.7
Mississippi Bluffs	47	–	–	–	4	7,900	10.9	–	–	–	29	90,175	13.6
Red River	28	5	7,075	15.5	3	31,548	17.1	–	–	–	8	63,500	11.4
Arkansas River	29	12	77,109	10.0	–	–	–	–	–	–	11	47,600	24.2

[a]Cropper and other share tenant majority.
[b]Renter majority.
[c]Seven additional plantations reported merchant loans but the amounts of the loans were not available.
[d]Sixteen additional plantations reported bank loans but the amounts of the loans were not available.

Table 26—LOANS CLOSED THROUGH FEDERAL INTERMEDIATE CREDIT BANKS FROM ORGANIZATION OF PRODUCTION CREDIT ASSOCIATIONS THROUGH DECEMBER 31, 1934

State	Number of Associations	Loans Closed		Average Loan
		Number	Amount	
United States	597	131,621	$92,882,001	$706
Seven cotton States	147	48,301	17,136,611	355
Seven cotton States as percent of United States	24.6	36.7	18.4	
Alabama	8	2,840	$ 985,170	347
Arkansas	30	6,925	2,581,103	373
Georgia	34	6,328	1,918,506	303
Louisiana	8	4,142	2,447,740	591
Mississippi	10	5,631	3,299,960	586
North Carolina	32	11,883	3,089,544	260
South Carolina	25	10,552	2,814,588	267

Source: *Farm Credit Administration, Second Annual Report, 1934*, Appendix Tables 43 and 54.

Table 27—NUMBER AND AMOUNT OF EMERGENCY CROP PRODUCTION LOANS IN SEVEN SOUTHEASTERN COTTON STATES, 1921-1930, 1931, 1932, 1933, AND 1934

State	Number					Amount (in thousands of dollars)				
	1921-1930	1931	1932	1933	1934	1921-1930	1931	1932	1933	1934
United States	121,330	438,932	507,632	633,585	445,198	$15,194	$55,787	$64,205	$57,376	$37,892
Seven cotton States	71,672	197,117	250,899	365,209	214,132	8,909	25,174	25,328	30,711	13,922
Seven cotton States as percent of United States	59.1	44.9	49.4	57.6	48.1	58.6	45.1	39.4	53.5	36.7
Alabama	20,714	19,738	19,658	33,723	15,253	$ 2,539	$ 2,663	$ 1,620	$ 2,352	$ 799
Arkansas	-	78,392	46,834	51,439	42,150	-	9,206	4,007	3,676	1,935
Georgia	28,680	17,775	44,158	63,147	39,207	3,585	2,521	4,887	5,517	3,034
Louisiana	-	24,853	26,185	35,896	22,678	-	3,197	2,416	2,625	1,221
Mississippi	-	30,790	40,065	56,346	27,942	-	4,442	3,890	4,244	1,474
North Carolina	1,711	17,705	36,742	64,051	35,097	214	2,182	4,181	6,014	2,702
South Carolina	20,567	7,864	37,257	60,607	31,805	2,571	963	4,327	6,283	2,757

Source: *Farm Credit Administration, Annual Report, 1933*, Appendix Table 35, and *Annual Report, 1934*, Appendix Table 64.

Table 28—SEED LOANS IN SEVEN SOUTHEASTERN COTTON STATES
(Appropriations of 1930-1934[a])

State	Year	Number of Loans	Number of Farms in Counties Included	Ratio of Loans to Farms (Percent)
Alabama	1930	6,438	174,211	3.7
	1931	19,738	255,154	7.7
	1932	20,218	257,395	7.9
	1933	33,723	257,395	13.1
	1934	15,272	257,395	5.9
Arkansas	1931	78,386	242,334	32.3
	1932	47,467	242,334	19.6
	1933	51,439	242,334	21.2
	1934	26,365	242,334	10.9
Georgia	1930	11,703	221,277	5.3
	1931	17,775	245,935	7.2
	1932	44,215	256,246	17.3
	1933	63,146	255,598	24.7
	1934	38,907	255,598	15.2
Louisiana	1931	24,853	122,280	20.3
	1932	26,411	154,661	17.1
	1933	35,897	161,445	22.2
	1934	21,613	161,445	13.4
Mississippi	1931	30,789	312,663	9.8
	1932	40,538	312,663	13.0
	1933	56,345	312,663	18.0
	1934	27,793	312,663	8.9
North Carolina	1931	17,706	207,384	8.5
	1932	36,955	279,698	13.2
	1933	64,051	279,708	22.9
	1934	34,851	279,708	12.5
South Carolina	1931	7,864	157,931	5.0
	1932	37,327	157,961	23.6
	1933	60,608	157,931	38.4
	1934	31,688	157,931	20.1

[a]1931 loans were made from the following appropriations:

Dec. 20, 1930	$45,000,000
Feb. 14, 1931	20,000,000
Feb. 23, 1931	2,000,000
Total	$67,000,000

Source: *Farm Credit Administration, Annual Reports for 1933 and 1934.*

Table 29—CASH COLLECTIONS MADE ON EMERGENCY CROP PRODUCTION AND FEED LOANS IN SEVEN SOUTHEASTERN COTTON STATES, 1931-1934

State	Amount Collected (in thousands of dollars)				Percent Collected of Value of Loans Made			
	1931[a]	1932[a]	1933[b]	1934[c]	1931[a]	1932[a]	1933[b]	1934[c]
United States	$25,135	$27,878	$34,133	$17,610	45.1	43.4	59.5	46.5
Seven cotton States	11,651	13,624	24,472	12,210	46.3	53.8	79.7	87.7
Value of collections in seven cotton States as percent of U.S. total collections	46.4	48.9	71.7	69.3				
Alabama	$ 818	$ 546	$ 1,896	$ 750	30.7	33.7	80.6	93.9
Arkansas	4,505	2,025	2,552	1,298	48.9	50.5	69.4	67.1
Georgia	939	1,806	4,830	2,821	36.9	37.0	87.6	93.0
Louisiana	1,320	1,637	1,883	1,050	41.3	67.7	71.7	86.0
Mississippi	2,293	2,239	2,921	1,300	51.6	57.6	68.8	88.2
North Carolina	1,346	2,923	4,942	2,443	61.7	69.9	82.2	90.0
South Carolina	430	2,448	5,448	2,548	44.6	56.6	86.7	92.4

[a]Collections through November 30, 1933.

[b]As of November 30, 1933. Only cash collections credited to principal on loans maturing by that date are included.

[c]As of December 31, 1934.

Source: *Farm Credit Administration, Annual Report, 1933,* Appendix Table 35, and *Annual Report, 1934,* Appendix Table 64.

Table 30—FINANCIAL RESULTS OF PLANTATION FAMILIES, BY TENURE STATUS, 1930-1933
(Cotton Plantation Enumeration)

Financial Result and Year	Total		Wage Hands		Croppers		Other Share Tenants		Renters		Displaced Tenants	
	Number	Percent	Number	Percent	Number	Percent	Number	Percent	Number	Percent	Number	Percent
1933												
Total	4,227	100.0	817	100.0	2,230	100.0	525	100.0	619	100.0	36	100.0
Gained	3,628	85.8	793	97.1	1,933	86.7	430	81.9	440	71.1	32	88.8
Lost	307	7.3	4	0.5	145	6.5	64	12.2	92	14.9	2	5.6
Even	292	6.9	20	2.4	152	6.8	31	5.9	87	14.0	2	5.6
1932												
Total	4,047	100.0	766	100.0	2,130	100.0	506	100.0	611	100.0	34	100.0
Gained	3,224	79.6	731	95.4	1,684	79.0	369	73.0	411	67.2	29	85.3
Lost	355	8.8	7	0.9	202	9.5	65	12.8	78	12.8	3	8.8
Even	468	11.6	28	3.7	244	11.5	72	14.2	122	20.0	2	5.9
1931												
Total	3,876	100.0	715	100.0	2,036	100.0	487	100.0	604	100.0	34	100.0
Gained	2,984	77.0	680	95.1	1,550	76.1	320	65.7	406	67.2	28	82.3
Lost	395	10.2	12	1.7	231	11.3	81	16.6	69	11.4	2	5.9
Even	497	12.8	23	3.2	255	12.6	86	17.7	129	21.4	4	11.8
1930												
Total	3,706	100.0	651	100.0	1,959	100.0	474	100.0	589	100.0	33	100.0
Gained	2,749	74.2	611	93.8	1,432	73.1	292	61.6	385	65.4	29	87.9
Lost	470	12.7	14	2.2	274	14.0	105	22.2	76	12.9	1	3.0
Even	487	13.1	26	4.0	253	12.9	77	16.2	128	21.7	3	9.1

Table 31—AMOUNT AND COST OF CREDIT USED BY 588 CROPPERS ON 112 FARMS IN COASTAL PLAIN REGION (NORTH CAROLINA), 1928

Type and Source of Credit	Total Amount of Credit	Average Amount of Credit		Actual Interest and Time Charges per Cropper	Weighted Average Term of Loan (months)	Weighted Average Cost of Credit per Annum (%)	Weighted Average Flat Rate per Dollar (%)
		Per Farm	Per Cropper				
Cash advances by farm owner	$ 64,053	$ 572	$109	$ 9	4.88	20.94	8.52
Farm supplies, fertilizer, etc., by farm owner and merchant	77,345	691	132	29	8.32	32.06	22.23
Household supplies by farm owner	66,718	596	113	25	4.82	53.46	21.49
Household supplies by merchant on farm owner's guarantee	31,768	284	54	15	4.70	71.29	27.94
Total supplies or merchant credit	175,831	1,570	299	69	6.34	43.50	22.98
Total advances, cash and supplies	239,884	2,142	408	78	5.95	38.56	19.12

Source: Wooten, H. H., *Credit Problems of North Carolina Cropper Farmers*, North Carolina Agricultural Experiment Station Bulletin 271, 1930, p. 14.

Table 32—GROSS INCOME[a] OF 645 PLANTATIONS, 1934
(Cotton Plantation Enumeration)

Size of Gross Income	Total Plantations Reporting[b]	
	Number	Percent
Total	645	100.0
Less than $2,000	96	14.9
$2,000 to 5,000	207	32.1
5,000 to 8,000	108	16.7
8,000 to 11,000	61	9.5
11,000 to 14,000	49	7.6
14,000 to 17,000	31	4.8
17,000 to 20,000	16	2.5
20,000 to 23,000	17	2.6
23,000 to 26,000	13	2.0
26,000 to 29,000	8	1.2
29,000 to 32,000	6	0.9
32,000 to 35,000	8	1.2
35,000 to 38,000	3	0.5
38,000 to 41,000	5	0.8
41,000 to 44,000	1	0.2
44,000 to 47,000	3	0.5
47,000 and over	13	2.0

[a]Cash and home use income of operator, supervisory employees, and tenants except renters and displaced tenants.
[b]Data not available for one plantation.

Table 33—NET INCOME FOR THE ONE-FOURTH OF THE PLANTATIONS IN EACH AREA WITH THE HIGHEST AND THE LOWEST NET INCOME PER PLANTATION, 1934
(Cotton Plantation Enumeration)

Area	Number of Plantations in Each Income Group	Net Income for One-fourth of Plantations in Each Area with Highest Net Income per Plantation			Net Income for One-fourth of Plantations in Each Area with Lowest Net Income per Plantation		
		Per Plantation	Per Capita	Per Crop Acre	Per Plantation	Per Capita	Per Crop Acre
Total	162	$14,010	$132	$20.42	$1,492	$ 68	$ 6.94
Atlantic Coast Plain	14	11,370	199	38.91	1,227	49	4.65
Upper Piedmont	10	6,788	148	22.21	1,653	57	10.21
Black Belt (A)[a]	28	6,467	112	14.73	855	48	4.08
Black Belt (B)[b]	25	5,415	104	16.23	552	69	2.07
Upper Delta	33	25,747	144	23.69	2,354	67	10.59
Lower Delta	12	9,577	120	22.44	231	46	1.94
Muscle Shoals	6	5,660	135	20.82	1,328	111	7.96
Interior Plain	8	19,354	136	16.19	1,793	66	11.97
Mississippi Bluffs	12	14,207	124	17.46	1,335	67	10.58
Red River	7	20,320	115	20.13	2,133	85	7.80
Arkansas River	7	37,140	121	18.15	5,589	119	12.20

[a]Cropper and other share tenant majority.
[b]Renter majority.

Table 34—SIZE OF PLANTATIONS IN RELATION TO PLANTATION NET INCOME, 1934
(Cotton Plantation Enumeration)

Area	Total Acres per Plantation			Crop Acres per Plantation			Number of Persons per Plantation		
	Average[a]	High 25%[b]	Low 25%[c]	Average[a]	High 25%[b]	Low 25%[c]	Average[a]	High 25%[b]	Low 25%[c]
Total	607	1,542	636	385	686	215	55	106	22
Atlantic Coast Plain	785	591	925	294	292	263	42	57	25
Upper Piedmont	437	544	395	211	306	162	33	46	29
Black Belt (A)[d]	785	1,473	482	275	447	187	34	62	18
Black Belt (B)[e]	840	820	1,040	256	334	266	27	52	8
Upper Delta	1,031	2,031	485	563	1,087	222	91	179	35
Lower Delta	1,146	2,328	730	207	444	119	35	80	5
Muscle Shoals	555	774	510	225	272	167	25	42	12
Interior Plain	1,160	2,848	391	523	1,195	150	58	142	27
Mississippi Bluffs	786	1,511	386	378	814	126	57	115	20
Red River	901	1,597	557	531	1,009	273	77	176	25
Arkansas River	1,722	3,324	1,031	998	2,046	458	143	308	47

[a]Based on 646 plantations.
[b]The term "High 25%" refers to the 25 percent of the plantations in each area with the highest net income per plantation.
[c]The term "Low 25%" refers to the 25 percent of the plantations in each area with the lowest net income per plantation.
[d]Cropper and other share tenant majority.
[e]Renter majority.

Table 35—SPECIALIZATION IN COTTON PRODUCTION IN RELATION TO PLANTATION NET INCOME, BY AREAS, 1934
(Cotton Plantation Enumeration)

Area	Cotton Acres per Plantation			Percentage of Crop Land in Cotton			Cotton Acres per Capita		
	Average[a]	High 25%[b]	Low 25%[c]	Average[a]	High 25%[b]	Low 25%[c]	Average[a]	High 25%[b]	Low 25%[c]
Total	151	306	56	39.4	44.5	26.1	2.77	2.87	2.52
Atlantic Coast Plain	80	71	53	26.9	24.5	20.2	1.90	1.24	2.16
Upper Piedmont	68	111	40	32.0	36.3	24.8	2.07	2.40	1.38
Black Belt (A)[d]	65	112	31	23.6	25.0	16.7	1.89	1.93	1.74
Black Belt (B)[e]	49	110	9	20.1	32.9	3.4	1.81	2.09	1.10
Upper Delta	265	516	99	47.1	47.5	44.7	2.86	2.88	2.85
Lower Delta	67	197	2	32.0	44.4	1.7	1.93	2.46	0.40
Muscle Shoals	67	124	31	29.6	45.4	18.8	2.64	2.96	2.54
Interior Plain	227	500	68	43.4	41.9	45.6	3.91	3.53	2.48
Mississippi Bluffs	157	380	36	41.4	46.7	28.5	2.76	3.31	1.77
Red River	283	550	125	53.3	54.5	45.8	3.68	3.12	4.97
Arkansas River	555	1,279	219	55.6	62.5	47.7	3.89	4.15	4.69

[a]Based on 646 plantations.
[b]The term "High 25%" refers to the 25 percent of the plantations in each area with the highest net income per plantation.
[c]The term "Low 25%" refers to the 25 percent of the plantations in each area with the lowest net income per plantation.
[d]Cropper and other share tenant majority.
[e]Renter majority.

Table 36—OPERATOR NET INCOME IN RELATION TO CAPITAL INVESTED, BY AREAS, 1934
(Cotton Plantation Enumeration)

Area	Total Plantations Reporting[a]	Operator's Capital Investment		Operator's Net Income			Percent on Investment, if Operator Is Arbitrarily Allowed for His Labor:		Annual Return for Operator's Labor if He Is Arbitrarily Allowed 6% per Annum on Investment
		Total	Average	Total	Average	Percent on Investment	$500 per Year	$1,000 per Year	
Total	632	$18,134,391	$28,694	$1,628,209	$2,576	9.0	7.2	5.5	$ 855
Atlantic Coast Plain	56	1,040,828	18,586	148,586	2,653	14.3	11.6	9.0	1,538
Upper Piedmont	39	514,300	13,187	67,735	1,737	13.2	9.4	5.6	946
Black Belt (A)[b]	111	1,985,443	17,887	163,639	1,474	8.2	5.4	2.6	401
Black Belt (B)[c]	93	1,372,640	14,760	124,987	1,344	9.1	5.7	2.3	459
Upper Delta	132	5,642,373	42,745	514,432	3,897	9.1	7.9	6.8	1,332
Lower Delta	49	873,715	17,831	84,677	1,728	9.7	6.9	4.1	658
Muscle Shoals	22	398,505	18,114	29,484	1,340	7.4	4.6	1.9	253
Interior Plain	30	1,395,479	46,516	70,957	2,365	5.1	4.0	2.9	– 426
Mississippi Bluffs	46	1,003,763	21,821	111,970	2,434	11.2	8.9	6.6	1,125
Red River	28	1,697,045	60,609	125,532	4,483	7.4	6.6	5.7	846
Arkansas River	26	2,210,300	85,012	186,210	7,162	8.4	7.8	7.2	2,061

[a]Data not available for 14 plantations.
[b]Cropper and other share tenant majority.
[c]Renter majority.

Table 37—OPERATOR LABOR INCOME[a] PER CROP ACRE AND PER ACRE VALUE OF LAND[b] 1934
(Cotton Plantation Enumeration)

Labor Income per Crop Acre	Total Plantations Reporting[c]		Per Acre Value of Land								Total Labor Income	Total Acres in Crops	Income per Crop Acre
	Number	Percent	Under $10	$10–$15	$15–$20	$20–$25	$25–$30	$30–$35	$35–$40	$40 and over			
Total	575	100.0	55	102	63	83	53	61	29	135	$449,014	223,261	$ 2.01
Loss $ 2.50 and over	87	15.1	8	10	7	8	7	5	4	38	-217,941	40,167	- 5.43
Loss less than $ 2.50	92	16.0	12	22	9	12	9	7	1	20	- 46,660	34,989	- 1.33
Gain less than $ 2.50	130	22.6	10	30	17	19	10	19	8	17	64,433	54,485	1.18
Gain $ 2.50 to $ 5.00	111	19.3	13	20	16	21	11	5	4	21	160,684	42,148	3.81
Gain $ 5.00 to $ 7.50	62	10.8	4	9	6	12	6	7	2	16	118,215	19,390	6.10
Gain $ 7.50 to $10.00	46	8.0	4	9	5	5	4	9	0	10	149,972	17,623	8.51
Gain $10.00 to $12.50	20	3.6	1	1	2	4	1	2	0	9	67,337	5,823	11.56
Gain $12.50 to $15.00	9	1.6	1	1	–	–	1	4	1	1	35,010	2,679	13.07
Gain $15.00 and over	18	3.1	2	–	1	2	4	3	3	3	117,964	5,957	19.80

[a]Labor income is net income (cash and home use) less 6 percent on capital investment in land, buildings (excluding operator's residence, gins, and commissaries), animals, and machinery. Enumerators were instructed to "enter values at conservative market value, not low assessed value or high speculative value."

[b]Land in crops, tillable land idle, pasture land, woods not pastured, and waste land.

[c]Data not available for 71 plantations; for 13 plantations operator's labor income data not available because value of rented and owned land not separated; for 58 plantations per acre value of land data not available because operators had no investment in land.

Table 38—NET INCOME BY TENURE STATUS, BY AREAS, 1934

(Cotton Plantation Enumeration)

Area	Total			Wage Hands			Croppers			Other Share Tenants			Renters		
	Total Families Reporting[a]	Net Income		Total Families Reporting	Net Income		Total Families Reporting	Net Income		Total Families Reporting	Net Income		Total Families Reporting	Net Income	
		Per Family	Per Capita		Per Family	Per Capita		Per Family	Per Capita		Per Family	Per Capita		Per Family	Per Capita
Total	5,093	$309	$73	865	$180	$62	2,873	$31[illegible]	$71	705	$417	$ 92	650	$354	$ 71
Atlantic Coast Plain	407	411	84	154	199	58	212	519	87	16	833	137	25	536	119
Upper Piedmont	245	326	73	53	153	68	124	336	66	52	440	82	16	444	109
Black Belt (A)[b]	754	311	67	175	156	52	404	334	66	62	313	64	113	471	83
Black Belt (B)[c]	679	256	54	174	175	56	232	267	50	23	408	63	250	289	57
Upper Delta	1,328	338	88	103	202	82	923	323	82	272	416	100	30	561	146
Lower Delta	346	199	47	16	205	76	136	154	38	49	217	52	145	234	50
Muscle Shoals	130	406	91	26	170	61	29	338	74	46	494	104	29	547	101
Interior Plain	227	371	83	7	70	61	172	334	74	48	546	114	-	-	-
Mississippi Bluffs	397	269	68	37	173	96	257	235	60	61	364	80	42	419	81
Red River	225	321	89	71	195	66	125	305	83	29	700	147	-	-	-
Arkansas River	355	245	76	49	213	69	259	243	75	47	290	87	-	-	-

[a]Data were not available for 29 families, including families of 2 wage hands, 13 croppers, 11 other share tenants, and 3 renters.

[b]Cropper and other share tenant majority.

[c]Renter majority.

Table 39—CROPPER AND OTHER SHARE TENANT NET INCOME, BY VALUE PER ACRE OF ALL PLANTATION LAND,[a] 1934
(Cotton Plantation Enumeration)

Value per Acre	Total			Croppers			Other Share Tenants		
	Number	Net Income		Number	Net Income		Number	Net Income	
		Total	Average		Total	Average		Total	Average
Total	3,585[b]	$1,192,133[c]	$333	2,880[d]	$898,192[e]	$312	705[f]	$293,941[g]	$417
Less than $10	252	72,565	288	205	55,418	270	47	17,147	365
$10 to 15	472	139,988	297	375	106,403	284	97	33,585	346
15 to 20	270	74,773	277	195	49,102	252	75	25,671	342
20 to 25	367	128,113	349	313	103,392	330	54	24,721	458
25 to 30	318	112,874	355	263	88,843	338	55	24,031	437
30 to 35	324	112,421	347	261	89,595	343	63	22,826	362
35 to 40	235	84,802	361	180	60,833	338	55	23,969	436
40 and over	985	350,465	356	776	247,794	319	209	102,671	491

[a]All land: crop, pasture, woods not pastured, and waste.
[b]Includes 362 families undistributed; excludes 17 families for which data were not available.
[c]Includes $116,132 undistributed.
[d]Includes 312 families undistributed; excludes 6 families for which data were not available.
[e]Includes $96,812 undistributed.
[f]Includes 50 families undistributed; excludes 11 families for which data were not available.
[g]Includes $19,320 undistributed.

Table 40—NET INCOME OF PLANTATION FAMILIES, BY INCOME FROM HOME USE PRODUCTION, 1934
(Cotton Plantation Enumeration)

Income from Home Use Production	Total Families Reporting[a]	Net Income												Median Net Income
		Loss $50 and Over	Loss Less Than $50	Gain Less Than $50	Gain $50–100	Gain $100–150	Gain $150–200	Gain $200–250	Gain $250–300	Gain $300–400	Gain $400–500	Gain $500–600	Gain $600 and Over	
Total	5,133	3	9	68	324	571	732	755	625	848	518	287	393	259
No home use income	364	–	1	13	105	133	66	24	10	3	5	2	2	124
Less than $40	1,115	1	3	43	131	225	290	195	101	92	21	6	7	177
$ 40 to 60	572	–	1	8	50	65	119	140	83	69	22	5	10	215
60 to 80	483	–	2	1	18	68	77	112	62	82	43	10	8	234
80 to 100	433	–	–	3	11	44	77	79	84	77	32	15	11	252
100 to 120	372	1	–	–	5	19	49	69	67	92	44	15	11	282
120 to 140	325	–	–	–	2	14	26	64	60	99	33	16	11	297
140 to 160	305	–	1	–	2	3	18	37	54	95	56	25	14	339
160 to 180	249	–	1	–	–	–	5	15	51	67	59	31	20	378
180 to 200	181	1	–	–	–	–	1	13	28	63	34	23	18	375
200 and over	734	–	–	–	–	–	4	7	25	109	169	139	281	538

[a]Data not available for 38 plantation families.

Table 41—CROPPER AND OTHER SHARE TENANT NET INCOME[a] PER FAMILY, BY OPERATOR LABOR INCOME PER CROP ACRE, 1934

(Cotton Plantation Enumeration)

Tenant Net Income per Family	Total Plantations Reporting[b]		Operator Labor Income per Crop Acre									Total Net Income	Total Cropper and Other Share Tenant Families on Plantations Reporting	Average Net Income per Family
	Number	Percent	Loss $2.50 and Over	Loss Less Than $2.50	Gain Less Than $2.50	Gain $2.50 - $5.00	Gain $5.00 - $7.50	Gain $7.50 - $10.00	Gain $10.00 - $12.50	Gain $12.50 - $15.00	Gain $15.00 and Over			
Total	546	100.0	74	64	122	117	72	48	20	12	17	$2,140,770	6,511	$329
$ 50 to $150	37	6.8	8	3	9	8	7	-	1	-	1	29,843	236	126
150 to 200	39	7.2	4	9	9	7	4	5	1	-	-	77,924	420	186
200 to 250	99	18.1	18	11	25	22	10	7	2	4	-	323,849	1,383	234
250 to 300	88	16.1	13	8	20	25	13	5	3	-	1	310,889	1,123	277
300 to 400	135	24.7	17	22	24	28	19	15	5	3	2	611,417	1,810	338
400 to 500	70	12.8	4	6	19	17	13	3	4	2	2	376,913	908	415
500 and over	78	14.3	10	5	16	10	6	13	4	3	11	409,935	631	650

[a]Cash and home use.

[b]Plantations operated entirely with renters and plantations operated entirely with non-resident wage hands are omitted; in addition 13 plantations are omitted because value of owned and rented land could not be separated.

Table 42—VALUE OF FARM DWELLINGS IN SEVEN SOUTHEASTERN COTTON STATES AS COMPARED TO GEOGRAPHIC DIVISIONS, 1930

Division and State	Mean Value (All Dwellings)	Median Value	
		Farms Operated by Owners Including Managers	Farms Operated by Tenants
United States	$1,126	$1,135	$ 472
New England	2,218	1,832	1,613
Middle Atlantic	2,237	1,986	2,058
East North Central	1,657	1,539	1,510
West North Central	1,559	1,521	1,247
South Atlantic	783	782	374
East South Central	503	512	314
West South Central	584	711	361
Mountain	989	806	682
Pacific	1,617	1,414	952
Seven cotton States	467	555	315
Alabama	408	499	297
Arkansas	391	495	301
Georgia	483	617	332
Louisiana	447	540	293
Mississippi	377	472	291
North Carolina	653	700	417
South Carolina	519	605	320

Source: *Fifteenth Census of the United States: 1930*, Agriculture Vol. II, Table 12, and Special Release.

Table 43—FARM HOUSES SURVEYED[a] IN SEVEN SOUTHEASTERN COTTON STATES, 1934

State	Total		White			Negro		
	Owners	Tenants	Total	Owners	Tenants	Total	Owners	Tenants
Seven cotton States	61,238	107,946	106,179	51,042	55,137	63,005	10,196	52,809
Alabama	8,138	16,644	16,300	6,702	9,598	8,482	1,436	7,046
Arkansas	7,837	12,528	13,553	6,613	6,940	6,812	1,224	5,588
Georgia	10,946	24,058	21,590	8,787	12,803	13,414	2,159	11,255
Louisiana	6,055	10,348	10,333	5,354	4,979	6,070	701	5,369
Mississippi	7,616	14,364	11,080	5,947	5,133	10,900	1,669	9,231
North Carolina	12,932	15,273	19,657	11,288	8,369	8,548	1,644	6,904
South Carolina	7,714	14,731	13,666	6,351	7,315	8,779	1,363	7,416

[a]Approximately 10 percent of all farms in each State were included in the survey.

Source: Farm Housing Survey by Bureau of Home Economics, U. S. Department of Agriculture, in cooperation with Civil Works Administration.

Table 44—PERCENT DISTRIBUTION OF FARM HOUSES SURVEYED IN SEVEN SOUTHEASTERN COTTON STATES, BY TYPE OF HOUSE AND BY COLOR AND TENURE OF OCCUPANTS, 1934

State	Percent							
	Total		White			Negro		
	Owners	Tenants	Total	Owners	Tenants	Total	Owners	Tenants
				Log				
Seven cotton States	3.5	3.4	3.8	3.4	4.2	2.9	4.1	2.7
Alabama	4.8	6.6	5.9	4.2	7.0	6.2	7.3	6.0
Arkansas	5.4	3.5	5.5	5.8	5.1	1.9	3.2	1.6
Georgia	3.3	3.1	3.6	3.3	3.9	2.3	3.2	2.2
Louisiana	2.3	1.2	1.8	2.1	1.4	1.3	3.4	1.1
Mississippi	3.5	2.6	3.1	3.2	3.0	2.8	4.8	2.4
North Carolina	3.3	3.8	3.7	3.2	4.4	3.4	4.5	3.1
South Carolina	1.9	2.4	2.2	1.9	2.5	2.3	2.1	2.3
				Earth				
Seven cotton States	*	*	*	*	*	0.1	0.2	0.1
Alabama	*	*	*	*	*	–	–	–
Arkansas	–	*	*	–	*	–	–	–
Georgia	–	–	–	–	–	–	–	–
Louisiana	*	0.1	0.1	0.1	0.2	*	–	*
Mississippi	0.3	*	*	*	*	0.2	1.1	*
North Carolina	–	*	*	–	*	–	–	–
South Carolina	0.1	0.2	*	0.1	*	0.4	–	0.5
				Frame (unpainted)				
Seven cotton States	56.5	79.5	62.0	51.6	71.5	86.6	80.5	87.7
Alabama	56.1	80.1	64.6	51.6	73.7	86.9	77.2	88.9
Arkansas	51.4	77.2	56.2	45.5	66.5	89.1	83.1	90.4
Georgia	58.5	80.7	64.3	52.1	72.6	89.1	84.6	89.9
Louisiana	62.7	87.5	70.5	60.3	81.3	91.8	80.8	93.0
Mississippi	65.2	86.2	69.5	60.3	80.0	88.4	82.5	89.6
North Carolina	48.9	67.6	52.1	45.3	61.5	74.8	73.4	75.2
South Carolina	57.6	79.1	62.4	52.5	70.7	86.3	81.0	87.3
				Frame (painted)				
Seven cotton States	39.1	16.8	33.5	43.9	23.9	10.3	15.0	9.4
Alabama	38.6	13.2	29.2	43.6	19.1	6.9	15.4	5.1
Arkansas	41.8	19.0	37.2	47.1	27.9	9.0	13.3	8.0
Georgia	37.7	16.1	31.7	43.9	23.4	8.6	12.2	7.9
Louisiana	34.9	11.2	27.6	37.4	17.1	6.9	15.8	5.7
Mississippi	30.2	11.0	26.6	35.6	16.3	8.5	11.5	8.0
North Carolina	45.9	28.0	42.6	49.4	33.3	21.5	21.8	21.4
South Carolina	39.5	18.0	34.7	44.5	26.3	10.9	16.6	9.8
				Frame (stucco)				
Seven cotton States	0.1	0.1	0.1	0.2	0.1	*	0.1	*
Alabama	*	*	*	*	*	*	–	*
Arkansas	0.2	0.1	0.2	0.2	0.1	*	0.2	–
Georgia	0.1	*	0.1	0.2	*	*	–	*
Louisiana	*	*	*	*	*	*	–	0.1
Mississippi	0.2	*	0.2	0.2	0.1	*	–	*
North Carolina	0.3	0.2	0.3	0.3	0.2	0.1	0.1	0.1
South Carolina	0.1	*	*	*	*	*	0.2	*
				Brick, Stone, and Concrete				
Seven cotton States	0.8	0.2	0.6	0.9	0.3	0.1	0.1	0.1
Alabama	0.5	0.1	0.3	0.6	0.2	*	0.1	*
Arkansas	1.2	0.2	0.9	1.4	0.4	*	0.2	*
Georgia	0.4	0.1	0.3	0.5	0.1	*	*	*
Louisiana	0.1	*	*	0.1	*	*	*	0.1
Mississippi	0.6	0.2	0.6	0.7	0.6	0.1	0.1	*
North Carolina	1.6	0.4	1.3	1.8	0.6	0.2	0.2	0.2
South Carolina	0.8	0.3	0.7	1.0	0.5	0.1	0.1	0.1

*Less than 0.05 percent.

Source: Farm Housing Survey by Bureau of Home Economics, U. S. Department of Agriculture, in cooperation with Civil Works Administration.

Table 45—NUMBER OF ROOMS PER FARM HOUSE SURVEYED IN SEVEN SOUTHEASTERN COTTON STATES, BY COLOR AND TENURE OF OCCUPANTS, 1934

State	Number of Rooms per House Surveyed							
	Total		White			Negro		
	Owners	Tenants	Total	Owners	Tenants	Total	Owners	Tenants
Seven cotton States	5.2	4.0	4.8	5.3	4.3	3.8	4.4	3.6
Alabama	4.9	3.6	4.4	5.0	4.0	3.4	4.0	3.2
Arkansas	4.5	3.8	4.2	4.6	3.9	3.7	3.9	3.7
Georgia	5.1	4.1	4.8	5.3	4.4	3.7	4.1	3.6
Louisiana	5.0	3.6	4.5	5.1	3.9	3.5	4.6	3.4
Mississippi	5.0	3.5	4.6	5.3	3.9	3.4	4.2	3.3
North Carolina	5.8	4.5	5.4	5.8	4.9	4.3	5.1	4.1
South Carolina	5.7	4.5	5.3	5.9	4.7	4.3	4.8	4.2

Source: Farm Housing Survey by Bureau of Home Economics, U. S. Department of Agriculture, in cooperation with Civil Works Administration.

Table 46—OCCUPANTS PER ROOM IN FARM HOUSES IN SEVEN SOUTHEASTERN COTTON STATES, BY COLOR AND TENURE OF OCCUPANTS, 1934

State	Number of Regular Occupants per Room							
	Total		White			Negro		
	Owners	Tenants	Total	Owners	Tenants	Total	Owners	Tenants
Seven cotton States	0.95	1.30	1.05	0.90	1.21	1.37	1.22	1.41
Alabama	1.00	1.49	1.15	0.97	1.30	1.72	1.36	1.81
Arkansas	0.98	1.29	1.09	0.95	1.25	1.31	1.18	1.34
Georgia	0.95	1.30	1.05	0.90	1.19	1.43	1.25	1.47
Louisiana	0.98	1.33	1.10	0.96	1.30	1.33	1.16	1.36
Mississippi	0.96	1.19	1.04	0.89	1.27	1.15	1.23	1.13
North Carolina	0.88	1.23	0.94	0.85	1.09	1.38	1.14	1.45
South Carolina	0.90	1.27	1.02	0.85	1.19	1.33	1.18	1.37

Source: Farm Housing Survey by Bureau of Home Economics, U. S. Department of Agriculture, in cooperation with Civil Works Administration.

Table 47—BEDROOMS AND OTHER ROOMS PER FARM HOUSE SURVEYED IN SEVEN SOUTHEASTERN COTTON STATES, BY COLOR AND TENURE OF OCCUPANTS, 1934

State	Total		White			Negro		
	Owners	Tenants	Total	Owners	Tenants	Total	Owners	Tenants
	Number of Bedrooms per House Surveyed							
Seven cotton States	2.9	2.3	2.7	2.9	2.4	2.2	2.6	2.1
Alabama	2.8	2.2	2.5	2.8	2.3	2.1	2.5	2.0
Arkansas	2.5	1.9	2.2	2.4	2.0	1.9	2.3	1.8
Georgia	2.9	2.4	2.8	3.0	2.6	2.3	2.6	2.2
Louisiana	2.7	2.1	2.6	2.7	2.3	2.0	2.8	1.9
Mississippi	2.9	2.0	2.6	2.9	2.3	2.0	2.6	1.9
North Carolina	3.2	2.5	2.9	3.2	2.6	2.4	2.7	2.4
South Carolina	3.1	2.7	2.9	3.2	2.7	2.7	2.8	2.7
	Number of Rooms Other than Bedrooms per House Surveyed							
Seven cotton States	2.3	1.7	2.1	2.4	1.9	1.6	1.8	1.5
Alabama	2.1	1.4	1.9	2.2	1.7	1.3	1.5	1.2
Arkansas	2.0	1.9	2.0	2.2	1.9	1.8	1.6	1.9
Georgia	2.2	1.7	2.0	2.3	1.8	1.4	1.5	1.4
Louisiana	2.3	1.5	1.9	2.4	1.6	1.5	1.8	1.5
Mississippi	2.1	1.5	2.0	2.4	1.6	1.4	1.6	1.4
North Carolina	2.6	2.0	2.5	2.6	2.3	1.9	2.4	1.7
South Carolina	2.6	1.8	2.4	2.7	2.0	1.6	2.0	1.5

Source: Farm Housing Survey by Bureau of Home Economics, U. S. Department of Agriculture, in cooperation with Civil Works Administration.

Table 48—PERCENT OF FARM HOUSES SURVEYED WITH SCREENS IN SEVEN SOUTHEASTERN COTTON STATES, BY COLOR AND TENURE OF OCCUPANTS, 1934

State	Percent							
	Total		White			Negro		
	Owners	Tenants	Total	Owners	Tenants	Total	Owners	Tenants
Seven cotton States	60.9	30.2	55.3	68.0	43.4	17.9	25.2	16.6
Alabama	59.7	26.1	51.6	68.6	39.7	9.4	18.0	7.6
Arkansas	76.0	59.1	76.0	80.7	71.7	44.8	50.8	43.5
Georgia	50.3	21.0	43.8	59.6	32.9	8.2	12.6	7.4
Louisiana	55.3	30.8	53.7	59.4	47.5	16.2	23.7	15.3
Mississippi	63.2	29.7	58.6	73.7	41.1	23.8	25.9	23.4
North Carolina	64.6	33.7	59.8	69.2	47.1	20.3	32.7	17.5
South Carolina	57.9	22.1	49.2	65.9	34.7	11.4	20.5	9.7

Source: Farm Housing Survey by Bureau of Home Economics, U. S. Department of Agriculture, in cooperation with Civil Works Administration.

Table 49—PERCENT DISTRIBUTION OF FARM HOUSES SURVEYED IN SEVEN SOUTHEASTERN COTTON STATES, BY SOURCE OF WATER, BY COLOR AND TENURE OF OCCUPANTS, 1934

State	Percent							
	Total		White			Negro		
	Owners	Tenants	Total	Owners	Tenants	Total	Owners	Tenants
				Well, Drilled or Driven				
Seven cotton States	20.3	23.3	20.4	20.4	20.3	25.3	19.7	26.4
Alabama	15.6	11.3	16.0	18.3	14.4	6.3	3.0	6.9
Arkansas	34.6	46.7	38.2	34.0	42.1	53.5	37.7	56.9
Georgia	15.3	12.8	12.8	14.3	11.7	14.8	19.0	14.0
Louisiana	29.4	30.8	31.4	31.5	31.3	28.4	13.3	30.3
Mississippi	16.2	34.9	18.3	15.3	21.9	38.6	19.4	42.1
North Carolina	19.7	28.6	20.3	18.1	23.4	34.1	31.1	34.8
South Carolina	16.0	10.3	13.3	16.9	10.1	10.7	12.3	10.4
				Well, Dug or Bored				
Seven cotton States	63.7	57.9	63.3	64.6	62.1	54.6	59.6	53.6
Alabama	78.4	70.3	75.3	79.8	72.2	68.4	72.3	67.6
Arkansas	42.3	28.2	38.4	42.7	34.4	24.0	40.3	20.4
Georgia	71.3	72.1	74.6	73.3	75.5	67.3	62.7	68.1
Louisiana	54.9	47.2	50.2	53.2	46.9	49.9	68.2	47.5
Mississippi	66.4	43.6	64.8	70.1	58.6	38.0	53.3	35.3
North Carolina	56.9	53.8	56.5	58.1	54.4	52.2	49.3	53.0
South Carolina	75.1	72.1	74.1	75.3	73.1	71.6	74.3	71.1
				Spring				
Seven cotton States	10.6	9.9	10.2	10.4	9.9	10.2	11.9	9.9
Alabama	7.3	11.1	6.8	4.8	8.2	15.7	19.3	15.0
Arkansas	9.8	6.9	10.5	10.7	10.4	3.0	4.9	2.6
Georgia	9.2	9.7	9.4	9.6	9.3	9.7	7.5	10.1
Louisiana	1.9	4.8	1.2	1.3	1.2	7.9	6.4	8.1
Mississippi	6.7	6.9	5.6	4.7	6.6	8.1	13.8	7.1
North Carolina	23.3	14.7	22.4	24.3	19.8	10.2	16.7	8.6
South Carolina	6.6	13.1	7.9	5.5	10.0	15.5	11.8	16.2
				Cistern				
Seven cotton States	5.9	3.8	5.9	6.5	5.4	2.2	2.7	2.0
Alabama	1.1	0.7	0.8	0.9	0.8	0.9	2.3	0.6
Arkansas	12.5	6.3	12.0	14.2	9.9	2.0	3.1	1.7
Georgia	2.7	2.1	3.2	3.1	3.3	0.9	1.3	0.8
Louisiana	25.5	18.7	27.6	27.9	27.2	10.4	7.6	10.7
Mississippi	8.2	4.8	8.3	8.4	8.3	3.6	7.5	2.8
North Carolina	0.3	0.1	0.3	0.3	0.2	0.1	-	0.1
South Carolina	0.1	*	0.1	0.1	*	*	-	*
				Stream				
Seven cotton States	0.6	0.9	0.7	0.6	0.7	1.1	0.8	1.1
Alabama	0.3	0.6	0.4	0.3	0.5	0.7	0.3	0.8
Arkansas	0.7	0.9	0.8	0.5	1.1	0.9	1.7	0.7
Georgia	0.5	0.5	0.6	0.5	0.7	0.3	0.1	0.4
Louisiana	0.7	3.2	0.7	0.3	1.1	5.0	3.1	5.2
Mississippi	0.6	0.7	0.6	0.5	0.6	0.8	1.0	0.7
North Carolina	0.7	0.5	0.7	0.8	0.6	0.4	0.1	0.5
South Carolina	1.1	0.8	1.0	1.2	0.8	0.8	0.7	0.9

*Less than 0.05 percent.

Source: Farm Housing Survey by Bureau of Home Economics, U. S. Department of Agriculture, in cooperation with Civil Works Administration.

Table 50.—PERCENT DISTRIBUTION OF FARM HOUSES SURVEYED IN SEVEN SOUTHEASTERN COTTON STATES, BY TYPE OF SANITARY FACILITIES, BY COLOR AND TENURE OF OCCUPANTS, 1934

State	Percent							
	Total		White			Negro		
	Owners	Tenants	Total	Owners	Tenants	Total	Owners	Tenants
	Outdoor Toilet (Improved)							
Seven cotton States	7.7	3.2	6.3	8.5	4.2	2.4	3.6	2.2
Alabama	5.4	2.0	3.9	5.9	2.4	1.7	3.0	1.4
Arkansas	5.1	3.7	3.9	5.3	2.5	4.9	3.8	5.2
Georgia	2.4	0.8	1.8	2.8	1.1	0.4	0.6	0.4
Louisiana	18.0	10.2	17.0	19.6	14.1	6.4	5.8	6.5
Mississippi	4.6	1.2	3.7	5.2	1.8	1.1	2.3	0.9
North Carolina	14.9	7.1	13.1	15.7	9.7	5.0	9.6	3.9
South Carolina	3.3	1.1	2.6	3.7	1.7	0.6	1.6	0.4
	Outdoor Toilet (Unimproved)							
Seven cotton States	66.7	67.4	67.3	66.6	68.0	66.8	67.2	66.7
Alabama	66.7	51.2	62.0	68.3	57.6	45.4	59.5	42.5
Arkansas	66.7	68.0	66.3	67.1	65.6	69.9	65.0	71.0
Georgia	80.0	81.6	82.5	79.7	84.4	78.9	80.8	78.5
Louisiana	66.7	67.9	65.7	65.2	66.2	70.5	78.2	69.5
Mississippi	67.8	73.1	72.8	71.2	74.7	69.7	55.6	72.2
North Carolina	51.7	54.1	52.2	51.0	53.8	54.9	56.9	54.4
South Carolina	72.1	69.6	69.6	71.1	68.2	71.9	76.7	71.1
	Indoor Toilet (Chemical)							
Seven cotton States	0.1	0.1	0.1	0.1	0.1	0.1	0.2	0.1
Alabama	*	*	*	*	–	*	–	*
Arkansas	0.1	0.4	0.4	0.2	0.6	–	–	–
Georgia	0.1	*	0.1	0.1	*	*	0.1	–
Louisiana	0.1	*	0.1	0.1	*	–	–	–
Mississippi	*	0.1	0.1	0.1	*	0.1	–	0.1
North Carolina	0.2	*	0.1	0.1	0.1	0.2	1.2	–
South Carolina	0.1	0.4	*	0.1	*	0.6	–	0.8
	Indoor Toilet (Flush)							
Seven cotton States	3.9	0.6	2.7	4.6	1.0	0.1	0.2	0.1
Alabama	2.2	0.2	1.2	2.6	0.3	*	0.2	*
Arkansas	2.3	0.6	1.8	2.6	1.0	0.1	0.2	0.1
Georgia	3.7	0.4	2.3	4.6	0.7	0.1	0.2	*
Louisiana	4.2	0.6	3.0	4.6	1.2	0.1	0.7	0.1
Mississippi	3.5	0.8	3.3	4.4	2.1	0.1	0.2	0.1
North Carolina	5.1	1.1	4.2	5.8	1.9	0.1	0.1	0.1
South Carolina	5.4	0.5	3.5	6.6	0.9	*	0.1	*
	No Sanitary Facilities Reported							
Seven cotton States	21.6	28.7	23.6	20.2	26.7	30.6	28.8	30.9
Alabama	25.7	46.6	32.9	23.2	39.7	52.9	37.3	56.1
Arkansas	25.8	27.3	27.6	24.8	30.3	25.1	31.0	23.7
Georgia	13.8	17.2	13.3	12.8	13.8	20.6	18.3	21.1
Louisiana	11.0	21.3	14.2	10.5	18.5	23.0	15.3	23.9
Mississippi	24.1	24.8	20.1	19.1	21.4	29.0	41.9	26.7
North Carolina	28.1	37.7	30.4	27.4	34.5	39.8	32.2	41.6
South Carolina	19.1	28.4	24.3	18.5	29.2	26.9	21.6	27.7

*Less than 0.05 percent.

Source: Farm Housing Survey by Bureau of Home Economics, U. S. Department of Agriculture, in cooperation with Civil Works Administration.

Table 51—PERCENT DISTRIBUTION OF FARM HOUSES SURVEYED IN SEVEN SOUTHEASTERN COTTON STATES, BY TYPE OF COOKING FACILITIES, BY COLOR AND TENURE OF OCCUPANTS, 1934

State	Percent							
	Total		White			Negro		
	Owners	Tenants	Total	Owners	Tenants	Total	Owners	Tenants
	Wood or Coal Stove							
Seven cotton States	94.9	93.7	95.7	94.7	96.6	91.4	95.6	90.6
Alabama	96.5	94.4	97.1	96.3	97.7	91.3	97.4	90.0
Arkansas	96.0	95.4	96.1	95.5	96.6	94.7	98.5	93.9
Georgia	97.2	96.5	98.0	97.4	98.4	94.6	96.2	94.3
Louisiana	91.1	93.9	90.1	90.1	90.0	97.7	98.9	97.6
Mississippi	93.2	79.5	95.8	95.2	96.5	72.5	86.1	70.1
North Carolina	93.1	96.4	93.7	92.5	95.5	97.5	97.2	97.6
South Carolina	96.5	97.4	96.9	96.2	97.6	97.3	98.0	97.2
	Kerosene or Gasoline Stove							
Seven cotton States	7.6	3.1	6.9	9.0	4.9	1.2	0.8	1.2
Alabama	3.6	1.1	2.6	4.3	1.5	0.6	0.5	0.6.
Arkansas	8.8	3.6	8.2	10.3	6.2	0.5	0.5	0.5
Georgia	7.2	1.7	5.3	8.8	2.9	0.3	0.5	0.2
Louisiana	15.7	7.9	16.8	17.6	16.0	0.5	1.1	0.4
Mississippi	3.8	3.8	3.6	4.7	2.4	4.0	0.6	4.6
North Carolina	9.4	4.6	9.2	10.5	7.4	1.5	2.1	1.3
South Carolina	5.9	1.8	5.0	7.1	3.2	0.4	0.6	0.3
	Gas Stove							
Seven cotton States	0.3	*	0.2	0.3	0.1	*	*	*
Alabama	0.1	*	0.1	0.1	*	–	–	–
Arkansas	0.4	0.1	0.3	0.5	0.2	*	0.1	*
Georgia	0.3	*	0.2	0.4	0.1	–	–	–
Louisiana	0.6	*	0.4	0.7	0.1	–	–	–
Mississippi	0.3	*	0.2	0.4	*	–	–	–
North Carolina	0.2	0.1	0.2	0.2	0.2	–	–	–
South Carolina	0.2	*	0.1	0.2	*	–	–	–
	Electric Stove							
Seven cotton States	0.7	0.1	0.5	0.9	0.2	*	*	*
Alabama	0.5	0.1	0.3	0.6	0.1	*	0.1	–
Arkansas	0.3	*	0.2	0.4	0.1	–	–	–
Georgia	0.8	0.1	0.4	1.0	0.1	*	–	*
Louisiana	0.6	*	0.4	0.7	0.1	–	–	–
Mississippi	0.1	*	0.1	0.2	*	*	–	*
North Carolina	1.0	0.3	0.9	1.2	0.5	*	–	*
South Carolina	1.5	0.1	0.9	1.8	0.2	–	–	–

*Less than 0.05 percent.

Source: Farm Housing Survey by Bureau of Home Economics, U. S. Department of Agriculture, in cooperation with Civil Works Administration.

Table 52—RURAL DEATHS FROM TYPHOID AND PARATYPHOID, PELLAGRA, AND MALARIA, IN SEVEN SOUTHEASTERN COTTON STATES AND IN OTHER STATES IN THE REGISTRATION AREA OF THE UNITED STATES, 1930

State	1930 Rural Population	Typhoid and Paratyphoid		Pellagra		Malaria	
		Deaths	Rate per 100,000	Deaths	Rate per 100,000	Deaths	Rate per 100,000
Seven cotton States	12,404,000	1,598	12.9	3,126	25.2	2,050	16.5
Alabama	1,904,800	152	8.0	445	23.4	267	14.0
Arkansas	1,472,400	283	19.2	312	21.2	567	38.5
Georgia	2,241,000	424	18.9	514	22.9	406	18.1
Louisiana	1,271,000	157	12.4	121	9.5	131	10.3
Mississippi	1,776,400	220	12.4	473	26.6	312	17.6
North Carolina	2,369,600	123	5.2	739	31.2	42	1.8
South Carolina	1,368,800	239	17.5	522	38.1	325	23.7
Other States in the registration area	48,160,600	2,415	5.0	1,325	2.8	939	1.9

Source: *Mortality Statistics 1930*, U. S. Bureau of the Census, Tables IA and 6.

Table 53—OCCUPATIONAL HISTORY OF 1,830 SOUTH CAROLINA FARMERS, BY 1933 TENURE AND BY COLOR

Color	Total Owners		Percent in Specified Combination - Owners (in 1933)								
	Number	Percent	Owner No Change	Tenant to Owner	Non-farming to Owner	Non-farming to Tenant to Owner	Tenant to Non-farming to Owner	Owner to Tenant to Owner	Hired Farm hand to Tenant to Owner	Tenant to Owner to Tenant to Owner	Other Combinations
White	515	100.0	27.1	34.9	11.4	3.3	2.5	0.8	2.7	1.4	15.9
Negro	162	100.0	14.2	39.5	6.2	4.3	0.6	1.2	12.4	-	21.6

Color	Total Tenants		Percent in Specified Combination - Tenants (in 1933)						
	Number	Percent	Tenant No Change	Non-farming to Tenant	Hired Farm Hand to Tenant	Tenant to Owner to Tenant	Tenant to Non-farming to Tenant	Owner to Tenant	Other Combinations
White	531	100.0	44.7	13.4	3.6	5.3	10.4	3.6	19.0
Negro	622	100.0	52.4	9.8	19.1	1.5	5.8	0.8	10.6

Note: The proportions of white and Negro owners that had been engaged in non-farming occupations at some time in life were 28.3 percent and 20.4 percent, respectively. Comparable figures for white and Negro tenants were 39.1 percent and 22.5 percent, respectively.

Source: Unpublished data from study conducted in eight representative farming counties by the South Carolina Experiment Station in cooperation with the C.W.A., E.R.A., and W.P.A.

Table 54—AGE OF NEGROES IN MICHIGAN, BY SEX, 1930

Age	Both Sexes		Male		Female	
	Number	Percent	Number	Percent	Number	Percent
All ages	169,453	100.0	88,936	100.0	80,517	100.0
Under 15 years	41,350	24.4	20,443	23.0	20,907	25.9
15 to 19 years	10,888	6.4	5,143	5.8	5,745	7.1
20 to 24 years	17,731	10.5	8,473	9.5	9,258	11.5
25 to 29 years	24,305	14.3	12,414	14.0	11,891	14.8
30 to 34 years	21,514	12.7	11,760	13.2	9,754	12.1
35 to 44 years	31,155	18.4	17,903	20.1	13,252	16.5
45 to 54 years	14,470	8.5	8,547	9.6	5,923	7.4
55 to 64 years	5,041	3.0	2,804	3.2	2,237	2.8
65 years and over	2,715	1.6	1,283	1.4	1,432	1.8
Unknown	284	0.2	166	0.2	118	0.1

Source: *Fifteenth Census of the United States: 1930*

Table 55—ILLITERACY IN THE POPULATION 10 YEARS OF AGE AND OVER, BY COLOR, AND 21 YEARS OF AGE AND OVER, BY SEX, 1930

State	10 Years of Age and Over (percent)						21 Years of Age and Over (percent)			
	Rank	Total	Rank	Native White	Rank	Negro	Rank	Male	Rank	Female
Iowa	1	0.8	7	0.4	2	2.0	1	1.0	1	0.9
Oregon	2	1.0	32	1.5	5	2.5	3	1.3	2	1.1
Washington	2	1.0	2	0.3	7	2.9	2	1.2	3	1.3
Idaho	4	1.1	7	0.4	17	4.2	6	1.5	3	1.3
South Dakota	5	1.2	7	0.4	4	2.2	4	1.4	9	1.9
Nebraska	5	1.2	7	0.4	13	3.9	4	1.4	7	1.6
Kansas	5	1.2	15	0.5	27	5.9	7	1.6	6	1.5
Utah	5	1.2	2	0.3	10	3.2	9	1.7	5	1.4
Minnesota	9	1.3	7	0.4	2	2.0	7	1.6	8	1.7
North Dakota	10	1.5	7	0.4	11	3.4	10	1.8	13	2.3
Wyoming	11	1.6	2	0.3	17	4.2	11	2.0	11	2.0
Indiana	12	1.7	27	0.9	28	6.0	13	2.3	9	1.9
Montana	12	1.7	2	0.3	21	4.6	12	2.1	13	2.3
Wisconsin	14	1.9	19	0.6	20	4.4	14	2.4	15	2.5
Michigan	15	2.0	22	0.7	8	3.0	14	2.4	17	2.7
Vermont	16	2.2	29	1.3	23	4.9	21	3.2	12	2.2
Missouri	17	2.3	32	1.5	32	8.8	19	3.1	15	2.5
Ohio	17	2.3	22	0.7	30	6.4	16	2.9	19	2.9
Illinois	19	2.4	19	0.6	12	3.6	16	2.9	23	3.3
California	20	2.6	2	0.3	9	3.1	18	3.0	21	3.2
Maine	21	2.7	34	1.6	22	4.8	23	3.9	17	2.7
New Hampshire	21	2.7	25	0.8	13	3.9	22	3.5	21	3.2
Oklahoma	23	2.8	35	1.7	33	9.3	23	3.9	20	3.1
Colorado	23	2.8	25	0.8	13	3.9	19	3.1	24	3.9
Pennsylvania	25	3.1	19	0.6	17	4.2	23	3.9	25	4.4
Massachusetts	26	3.5	7	0.4	26	5.4	27	4.0	28	5.0
New York	27	3.7	15	0.5	5	2.5	23	3.9	29	5.4
New Jersey	28	3.8	15	0.5	25	5.1	28	4.4	30	5.5
Maryland	28	3.8	29	1.3	36	11.4	29	5.0	25	4.4
Delaware	30	4.0	28	1.2	37	13.2	31	5.2	27	4.9
Nevada	31	4.5	1	0.2	1	1.5	29	5.0	32	5.6
Connecticut	31	4.5	7	0.4	23	4.9	31	5.2	33	6.8
West Virginia	33	4.8	40	3.7	34	11.3	34	7.0	30	5.5
Rhode Island	34	4.9	22	0.7	31	8.1	33	5.8	35	7.0
Kentucky	35	6.6	46	5.7	40	15.4	38	9.3	33	6.8
Arkansas	36	6.8	39	3.5	41	16.1	37	9.2	39	8.2
Texas	36	6.8	31	1.4	38	13.4	35	7.8	38	8.0
Florida	38	7.1	36	1.9	42	18.8	36	8.7	37	7.8
Tennessee	39	7.2	44	5.4	39	14.9	39	10.3	36	7.7
Virginia	40	8.7	40	3.7	34	11.3	41	12.1	40	9.5
Georgia	41	9.4	38	3.3	43	19.9	42	12.5	41	10.9
North Carolina	42	10.0	45	5.6	44	20.6	44	14.2	42	12.0
Arizona	43	10.1	15	0.5	16	4.0	40	10.4	43	13.7
Alabama	44	12.6	42	4.8	47	26.2	46	16.6	44	15.2
Mississippi	45	13.1	37	2.7	45	23.2	47	18.0	45	15.6
New Mexico	46	13.3	48	7.7	28	6.0	43	13.6	48	20.3
Louisiana	47	13.5	47	7.3	46	23.3	45	16.5	46	17.2
South Carolina	48	14.9	43	5.1	48	26.9	48	18.8	47	18.4

Source: *Fifteenth Census of the United States: 1930*, Population Vol. III, Part 1.

Table 56—RESIDENCE, COLOR, AND LOCATION IN COTTON OR NON-COTTON COUNTIES[a] OF RELIEF CASES IN SEVEN SOUTHEASTERN COTTON STATES,[b] OCTOBER 1933

Area	Total			Urban			Rural		
	Total	White	Negro	Total	White	Negro	Total	White	Negro
	Number Relief Cases								
All counties	493,244	275,178	218,066[c]	182,199	78,716	103,483	311,045	196,462	114,583
Cotton counties	283,052	160,641	122,411	80,149	36,300	43,849	202,903	124,341	78,562
Non-cotton counties	210,192	114,537	95,655	102,050	42,416	59,634	108,142	72,121	36,021
	Percentage Distribution by Residence								
All counties	100.0	100.0	100.0	36.9	28.6	47.5	63.1	71.4	52.5
Cotton counties	100.0	100.0	100.0	28.3	22.6	35.8	71.7	77.4	64.2
Non-cotton counties	100.0	100.0	100.0	48.6	37.0	62.3	51.4	63.0	37.7
	Percentage Distribution by Color								
All counties	100.0	55.8	44.2	100.0	43.2	56.8	100.0	63.2	36.8
Cotton counties	100.0	56.8	43.2	100.0	45.3	54.7	100.0	61.3	38.7
Non-cotton counties	100.0	54.5	45.5	100.0	41.6	58.4	100.0	66.7	33.3
	Percentage Distribution by Cotton and Non-cotton Counties								
All counties	100.0	100.0	100.0	100.0	100.0	100.0	100.0	100.0	100.0
Cotton counties	57.4	58.4	56.1	44.0	46.1	42.4	65.2	63.3	68.6
Non-cotton counties	42.6	41.6	43.9	56.0	53.9	57.6	34.8	36.7	31.4

[a] A cotton county is one in which 40 percent or more of the gross farm income in 1929 came from cotton farms. Of the 593 counties in the seven States, 335 or 56 percent were cotton counties in 1929. Data complete for 539 of the 593 counties; data not available for 45 all-rural counties; rural data not available for 8 other counties; urban data not available for 1 county.

[b] Alabama, Arkansas, Georgia, Louisiana, Mississippi, North Carolina, and South Carolina.

[c] Includes 353 cases of "other races."

Source: *Unemployment Relief Census, October 1933*, F.E.R.A., Report Number Two, Table 9, pp. 106–211.

Table 57—COMBINED RURAL RELIEF AND REHABILITATION INTENSITY RATES,[a] OCTOBER 1933 THROUGH JUNE 1935, FOR RURAL COUNTIES,[b] BY AGRICULTURAL AREAS

Year and Month[c]	All Areas	Eastern Cotton	Western Cotton	Appa-lachian-Ozark	Hay and Dairy	Lake States Cut-over	Corn Belt	Spring Wheat	Winter Wheat	Ranching
1933										
October	11.5	12.8	8.5	16.8	2.8	22.7	4.7	12.6	12.0	6.2
November	14.1	15.4	10.0	19.9	3.6	22.6	6.3	18.7	16.3	9.5
December	10.9	12.9	8.9	13.0	3.3	18.5	4.1	23.6	9.2	9.3
1934										
January	12.8	17.1	8.9	15.0	3.8	14.6	5.5	26.9	4.6	8.6
February	12.6	16.9	9.6	16.4	3.9	14.2	5.5	14.7	3.3	10.1
March	13.5	13.5	11.3	19.7	5.1	19.5	6.6	24.7	8.9	12.2
April	11.8	10.5	12.2	12.9	5.9	26.4	8.2	24.9	13.3	13.3
May	14.0	11.7	13.6	19.5	9.1	26.9	8.6	24.1	11.8	13.4
June	15.4	11.9	12.9	20.8	18.1	32.9	9.4	30.9	10.7	13.3
July	16.2	13.0	17.5	22.6	16.3	26.8	9.7	26.6	14.8	13.6
August	18.0	15.4	23.3	23.9	15.8	27.5	10.6	26.7	18.1	14.6
September	17.7	14.2	22.4	22.4	14.5	33.3	11.7	29.1	18.6	15.0
October	17.0	11.4	22.9	22.9	13.8	27.3	12.6	33.6	18.2	14.4
November	17.9	11.2	27.4	24.0	16.4	29.0	13.5	33.7	20.5	15.3
December	18.8	12.3	29.1	24.6	14.4	32.2	14.5	33.8	19.7	17.8
1935										
January	19.6	13.2	30.0	25.4	12.7	34.8	15.7	34.6	20.8	20.3
February	19.6	12.7	29.6	25.6	14.0	37.5	15.3	35.2	20.9	21.1
March	19.8	12.7	26.2	26.7	15.3	39.6	15.4	35.6	20.6	21.0
April	19.1	11.9	23.1	27.0	14.8	40.3	14.2	35.6	19.4	19.8
May	18.3	11.5	21.1	26.7	13.4	35.6	13.3	33.8	19.4	17.6
June	16.5	10.7	18.7	25.2	9.7	29.2	11.9	30.6	16.2	15.3

[a] Percentage ratio of total estimated number of relief and rehabilitation cases to all families of the same residence class in 1930. The rehabilitation program began in April 1934.

[b] A rural county is defined as one in which there was no place of 2,500 or more population in 1930.

[c] Occasional county reports were not available for scattered months but the small number of such omissions is not sufficient to invalidate area intensity rates.

Source: Division of Research, Statistics and Finance, Federal Emergency Relief Administration.

Table 58—RELIEF AND REHABILITATION CASES IN COTTON COUNTIES[a] IN SEVEN SOUTHEASTERN COTTON STATES[b] BY 6-MONTH INTERVALS, MAY 1933 THROUGH NOVEMBER 1935, BY STATES

State	Relief Cases						Rehabilitation Cases				Relief and Rehabilitation Cases					
	May 1933	Nov. 1933	May 1934	Nov. 1934	May 1935	Nov. 1935	May 1934	Nov. 1934	May 1935	Nov. 1935[c]	May 1933	Nov. 1933	May 1934	Nov. 1934	May 1935	Nov. 1935
Seven cotton States	391,126	292,643	250,902	233,465	195,469	94,682	13,353	30,424	55,757	39,534	391,126	292,643	264,255	263,899	251,226	134,216
Alabama	84,009	77,189	55,074	27,589	30,293	13,137	2,192	5,476	16,149	8,584	84,009	77,189	57,266	33,065	46,442	21,721
Arkansas	83,827	37,697	34,410	44,704	41,028	15,083	4,205	3,423	11,854	6,535	83,827	37,697	38,615	48,127	52,882	21,618
Georgia	29,129	25,812	21,537	27,858	22,434	9,179	24	995	4,500	4,333	29,129	25,812	21,561	28,853	26,934	13,512
Louisiana	24,168	27,153	12,892	9,614	10,064	5,161	6,760	16,801	6,052	4,327	24,168	27,153	19,652	26,415	16,116	9,488
Mississippi	55,463	42,220	53,570	48,739	36,615	23,776	d	3,017	10,990	10,457	55,463	42,220	53,570	51,756	47,605	34,233
North Carolina	30,666	18,977	17,491	21,294	16,255	12,245	d	d	1,737	1,775	30,666	18,977	17,491	21,294	17,992	14,020
South Carolina	83,864	63,595	55,928	53,667	38,780	16,101	172	712	4,475	3,523	83,864	63,595	56,100	54,379	43,255	19,624

[a] A cotton county is defined as one in which 40 percent or more of the 1929 gross farm income came from cotton farms. Data not available for 12 cotton counties.

[b] Alabama, Arkansas, Georgia, Louisiana, Mississippi, North Carolina, and South Carolina.

[c] The November 1935 rehabilitation load consisted of standard and emergency cases on the active rolls of the Resettlement Administration. Statistics were available only by States, the number of cases allocated to cotton counties being based upon the proportion of the May 1935 rehabilitation load in cotton counties.

[d] No rehabilitation program, or rehabilitation statistics not reported separately from relief statistics.

Source: Division of Research, Statistics and Finance, F.E.R.A., and Resettlement Administration.

Table 59—TENURE AND COLOR OF FARM OPERATORS IN SEVEN SOUTHEASTERN COTTON STATES,[a] 1930 and 1935

Tenure and Color	1930		1935		Number Increase or Decrease	Percent Increase or Decrease
	Number	Percent	Number	Percent		
Total	1,667,074	100.0	1,725,382	100.0	+ 58,308	+ 3.5
Full owners	503,892	30.2	562,582	32.6	+ 58,690	+11.6
Part owners	90,412	5.4	94,566	5.5	+ 4,154	+ 4.6
Managers	5,718	0.4	4,872	0.3	- 846	-14.8
All tenants	1,067,052	64.0	1,063,362	61.6	- 3,690	- 0.3
Croppers	543,773	32.6	513,765	29.8	- 30,008	- 5.5
Other tenants	523,279	31.4	549,597	31.8	+ 26,318	+ 5.0
Total white	995,921	100.0	1,104,291	100.0	+108,370	+10.9
Full owners	421,221	42.3	476,620	43.2	+ 55,399	+13.2
Part owners	65,760	6.6	72,011	6.5	+ 6,251	+ 9.5
Managers	5,383	0.5	4,720	0.4	- 663	-12.3
All tenants	503,557	50.6	550,940	49.9	+ 47,383	+ 9.4
Croppers	220,229	22.1	200,870	18.2	- 19,359	- 8.8
Other tenants	283,328	28.5	350,070	31.7	+ 66,742	+23.6
Total Negro	761,153	100.0	621,091	100.0	- 50,062	- 7.5
Full owners	82,671	12.3	85,962	13.9	+ 3,291	+ 4.0
Part owners	24,652	3.7	22,555	3.6	- 2,097	- 8.5
Managers	335	*	152	*	- 183	-54.6
All tenants	563,495	84.0	512,422	82.5	- 51,073	- 9.1
Croppers	323,544	48.2	312,895	50.4	- 10,649	- 3.3
Other tenants	239,951	35.8	199,527	32.1	- 40,424	-16.8

*Less than 0.05 percent.

[a]Alabama, Arkansas, Georgia, Louisiana, Mississippi, North Carolina, and South Carolina.

Source: *United States Census of Agriculture: 1935*, Preliminary Figures, and *United States Census of Agriculture: 1930*, Volume II, Part 2, County Table 1, Supplemental for the Southern States.

Table 60—CHANGE IN NUMBER OF TENANTS, BY TENURE AND BY COTTON AND TOBACCO COUNTIES[a] AND NON-COTTON AND NON-TOBACCO COUNTIES IN SEVEN SOUTHEASTERN COTTON STATES, 1930-1935

State	Increase or Decrease 1930 - 1935																	
	All Counties (593)						Cotton and Tobacco Counties (367)[a]						Non-cotton and Non-tobacco Counties (226)					
	All Tenants		Croppers		Other Tenants		All Tenants		Croppers		Other Tenants		All Tenants		Croppers		Other Tenants	
	Number	Percent	Number	Percent	Number	Percent	Number	Percent	Number	Percent	Number	Percent	Number	Percent	Number	Percent	Number	Percent
Seven cotton States	- 3,690	- 0.3	-30,008	- 5.5	+26,318	+ 5.0	-18,520	- 2.1	-22,798	- 5.0	+ 4,278	+ 1.0	+14,830	+ 7.9	- 7,210	- 8.6	+22,040	+21.1
Alabama	+ 9,827	+ 5.9	+ 2,840	+ 4.4	+ 6,987	+ 6.9	+ 7,151	+ 4.6	+ 3,029	+ 5.1	+ 4,122	+ 4.4	+ 2,676	+21.6	- 189	- 3.6	+ 2,865	+39.7
Arkansas	- 932	- 0.6	- 9,431	-12.6	+ 8,499	+10.9	- 4,431	- 3.5	- 8,763	-13.0	+ 4,332	+ 7.2	+ 3,499	+13.9	- 668	- 8.6	+ 4,167	+23.8
Georgia	-10,059	- 5.8	-20,429	-20.3	+10,370	+14.1	- 7,319	- 9.2	-11,360	-23.2	+ 4,041	+13.2	- 2,740	- 2.9	- 9,069	- 17.5	+ 6,329	+14.8
Louisiana	+ 826	+ 0.8	+ 791	+ 1.6	+ 35	+ 0.1	- 1,251	- 1.5	+ 709	+ 1.7	- 1,960	- 4.7	+ 2,077	+ 9.0	+ 82	+ 1.2	+ 1,995	+12.5
Mississippi	- 8,053	- 3.6	+ 1,620	+ 1.2	- 9,673	-10.7	- 8,958	- 4.1	+ 1,237	+ 0.9	-10,195	-11.6	+ 905	+19.0	+ 383	+ 14.5	+ 522	+24.8
North Carolina	+ 4,543	+ 3.3	- 2,698	- 3.9	+ 7,241	+10.6	- 2,119	- 1.9	- 4,592	- 7.7	+ 2,473	+ 4.8	+ 6,662	+25.7	+ 1,894	+ 20.7	+ 4,768	+28.4
South Carolina	+ 158	+ 0.2	- 2,701	- 5.5	+ 2,859	+ 5.3	- 1,593	- 1.6	- 3,058	- 6.3	+ 1,465	+ 2.8	+ 1,751	+88.7	+ 357	+315.9	+ 1,394	+74.9

[a]A cotton county is one in which 40 percent or more of the gross farm income in 1929 came from cotton farms. A tobacco county is one in which 10 percent or more of the cultivated acreage in 1929 was planted in tobacco.

Source: *United States Census of Agriculture: 1935.*

Table 61—PERCENT CHANGE IN THE NUMBER OF TENANTS IN SEVEN SOUTHEASTERN COTTON STATES,[a] BY COTTON AND TOBACCO COUNTIES[b] AND NON-COTTON AND NON-TOBACCO COUNTIES, 1930-1935

Percent Change 1930-1935	All Tenants			Croppers			Other Tenants		
	All Counties	Cotton and Tobacco Counties	Non-cotton and Non-tobacco Counties	All Counties	Cotton and tobacco Counties	Non-cotton and Non-tobacco Counties	All Counties	Cotton and Tobacco Counties	Non-cotton and Non-tobacco Counties
Total: Number	593	367	226	593	367	226	593	367	226
Total: Percent	100.0	100.0	100.0	100.0	100.0	100.0	100.0	100.0	100.0
Increase									
(+36) and over	10.3	2.5	23.0	14.5	9.8	22.1	16.2	8.7	28.3
(+26) - (+35)	4.7	3.8	6.2	4.4	4.1	4.9	10.3	7.4	15.0
(+16) - (+25)	9.6	7.9	12.4	6.2	6.8	5.3	15.9	15.2	16.8
(+ 6) - (+15)	16.7	17.7	15.1	11.6	10.1	14.2	19.9	21.3	17.7
Little change									
(+ 5) - (- 4)	28.5	32.7	21.7	13.5	15.8	9.8	17.7	20.4	13.3
Decrease									
(- 5) - (-14)	22.3	27.0	14.6	15.9	18.5	11.5	11.0	14.2	5.8
(-15) - (-24)	7.4	7.9	6.6	20.4	21.3	19.0	5.7	7.6	2.7
(-25) - (-34)	0.5	0.5	0.4	7.9	8.7	6.6	2.0	3.0	0.4
(-35) and over	-	-	-	5.6	4.9	6.6	1.3	2.2	-

[a]Alabama, Arkansas, Georgia, Louisiana, Mississippi, North Carolina, and South Carolina.
[b]A cotton county is one in which 40 percent or more of the gross farm income in 1929 came from cotton farms. A tobacco county is one in which 10 percent or more of the cultivated acreage in 1929 was planted in tobacco.
Source: *United States Census of Agriculture: 1935.*

Table 62—INCIDENCE OF RELIEF AMONG PLANTATION FAMILIES, 1934 TO JULY 1, 1935, BY MONTHS AND BY TENURE STATUS (Cotton Plantation Enumeration)

Year and Month on Relief[a]	Tenure Status of Families Receiving Relief						
	All Families Receiving Relief		Renters	Other Share Tenants	Croppers	Wage Hands	Displaced Tenants
	Number	Percent					
1934							
Total number of cases reporting[b]	5,033		642	703	2,781	860	47
Unduplicated total receiving relief	719	14.3	53	177	401	72	16
January	374	7.4	33	84	209	42	6
February	439	8.7	33	114	244	42	6
March	311	6.2	30	76	159	34	12
April	235	4.7	20	51	119	32	13
May	143	2.8	15	46	50	22	10
June	135	2.7	13	38	55	18	11
July	140	2.8	13	42	52	22	11
August	143	2.8	11	46	59	14	13
September	124	2.5	10	38	52	14	10
October	119	2.4	10	29	59	12	9
November	120	2.4	11	25	59	16	9
December	139	2.8	9	34	69	18	9
1935							
Total number of cases reporting[b]	5,031		642	702	2,780	860	47
Unduplicated total receiving relief	155	3.1	13	15	104	11	12
January	129	2.6	11	13	83	11	11
February	120	2.4	9	8	86	7	10
March	52	1.0	5	3	30	4	10
April	32	0.6	2	5	13	2	10
May	30	0.6	2	3	13	2	10
June	29	0.6	1	3	13	4	8

[a]Data by months were not available for all cases. Therefore, weighted figures are given for the number of cases on relief in a given month.
[b]For total families enumerated by tenure status, see Appendix B, Table E.

Table 63—OBLIGATIONS[a] INCURRED FOR EMERGENCY RELIEF, BY SOURCE OF FUNDS, JANUARY 1933 THROUGH SEPTEMBER 1935, IN UNITED STATES AND SEVEN SOUTHEASTERN COTTON STATES

State	Total			Federal Funds			State Funds			Local Funds		
	Amount	Per Capita	Percent	Amount	Per Capita	Percent	Amount	Per Capita	Percent	Amount	Per Capita	Percent
United States	$3,809,784,015	$31.03	100.0	$2,726,487,865	$22.21	71.6	$461,656,119	$3.76	12.1	$621,640,031	$5.06	16.3
Seven cotton States	287,331,042	17.49	100.0	276,908,586	16.85	96.4	631,156	.04	0.2	9,791,300	.60	3.4
Alabama	45,650,584	7.25	100.0	43,344,257	16.38	94.9	163,742	.06	0.4	2,142,585	.81	4.7
Arkansas	40,541,707	21.86	100.0	39,077,384	21.07	96.4	301,712	.16	0.7	1,162,611	.63	2.9
Georgia	46,344,221	15.93	100.0	44,036,621	15.14	95.0	5	b	*	2,307,595	.79	5.0
Louisiana	51,289,446	24.41	100.0	49,675,984	23.64	96.9	1,697	b	*	1,611,765	.77	3.1
Mississippi	30,201,995	15.03	100.0	29,418,788	14.64	97.4	164,000	.08	0.5	619,207	.31	2.1
North Carolina	37,818,222	11.92	100.0	36,566,565	11.53	96.7	–	–	–	1,251,657	.39	3.3
South Carolina	35,484,867	20.41	100.0	34,788,987	20.01	98.0	–	–	–	695,880	.40	2.0

*Less than 0.05 percent.

[a] Includes obligations incurred for relief extended under the general relief program and all special programs, and for administration.

[b] Less than one cent.

Source: *Monthly Report of the F.E.R.A.*, October 1935, Table 5, and *Fifteenth Census of the United States: 1930*.

Table 64—RELIEF GRANTS TO RURAL RELIEF HOUSEHOLDS, BY AREAS, JUNE 1935[a]
(138 Counties Representing 9 Agricultural Areas)

Area	Total Cases		Percent According to Size of Relief Grant							Median Size[b] of Relief Grant
	Number	Percent	$ 1 to $10	$10 to $20	$20 to $40	$40 to $60	$60 to $80	$ 80 to $100	$100 and over	
					Agricultural Cases					
All areas	23,394	100.0	38.8	37.7	19.0	3.7	0.7	0.1	–	$11.84
Eastern Cotton	3,308	100.0	53.2	37.2	7.4	1.8	0.4	–	–	8.98
Western Cotton	3,764	100.0	57.7	35.5	6.0	0.6	0.2	–	–	8.61
Appalachian-Ozark	6,622	100.0	48.0	43.1	8.0	0.8	0.1	–	–	9.81
Lake States Cut-over	952	100.0	16.8	32.2	35.7	11.3	2.9	1.1	–	19.93
Hay and Dairy	2,370	100.0	12.0	31.8	42.9	11.1	2.1	0.1	–	21.54
Corn Belt	2,778	100.0	26.2	35.9	32.6	4.5	0.5	0.2	0.1	15.95
Spring Wheat	2,212	100.0	20.9	34.7	33.8	8.2	1.8	0.3	0.3	17.76
Winter Wheat	666	100.0	37.3	35.1	21.6	6.0	–	–	–	12.26
Ranching	722	100.0	10.0	45.7	40.4	2.8	1.1	–	–	18.20
					Non-Agricultural Cases					
All areas	21,446	100.0	27.8	35.7	27.8	6.8	1.6	0.2	0.1	14.64
Eastern Cotton	2,152	100.0	31.9	46.4	15.6	5.1	1.0	–	–	12.44
Western Cotton	1,654	100.0	53.0	35.5	8.9	1.5	1.1	–	–	9.10
Appalachian-Ozark	7,046	100.0	35.8	40.5	20.5	2.6	0.5	0.1	–	12.15
Lake States Cut-over	1,858	100.0	18.7	28.3	35.6	12.4	3.8	1.0	1.2	20.96
Hay and Dairy	4,096	100.0	13.5	27.1	41.4	14.9	2.6	0.4	0.1	23.27
Corn Belt	3,090	100.0	22.2	33.3	37.1	6.3	0.8	0.3	–	17.65
Spring Wheat	586	100.0	13.7	26.6	40.6	11.9	6.5	–	0.7	22.74
Winter Wheat	396	100.0	29.8	32.8	29.8	5.6	2.0	–	–	15.33
Ranching	568	100.0	15.8	46.8	31.7	3.9	1.8	–	–	16.23

[a]Exclusive of cases opened, reopened, or closed during the month.
[b]Medians calculated from original table with smaller class intervals.
Source: Survey of Current Changes in the Rural Relief Population, Division of Social Research, W.P.A.

Table 65—CASES RECEIVING RURAL REHABILITATION ADVANCES DURING JUNE 1935, CASES EVER UNDER CARE, AND AMOUNT OF GOODS ISSUED UNDER THE ENTIRE RURAL REHABILITATION PROGRAM, IN UNITED STATES AND SEVEN SOUTHEASTERN COTTON STATES

State	Cases Receiving Advances June 1935	Total Cases Ever Under Care	All Goods Issued		Subsistence Goods		Capital Goods	
			Amount	Per Case[a]	Amount	Per Case[a]	Amount	Per Case[a]
United States	203,418	397,130	$49,039,382	$124	$14,117,634	$36	$34,921,748	$88
Seven cotton States	84,713	137,901	22,873,508	166	8,685,898	63	14,187,610	103
Alabama	17,507	31,923	3,351,201	105	1,603,495	50	1,747,706	55
Arkansas	18,998	25 489	5,424,138	213	2,143,667	84	3,280,471	129
Georgia	12,394	13,701	4,059,036	296	827,118	60	3,231,918	236
Louisiana	10,710	36,879	3,224,147	87	2,508,991	68	715,156	19
Mississippi	12,360	14,705	3,350,165	288	945,618	64	2,404,547	164
North Carolina	6,665	8,435	1,626,634	193	312,493	37	1,314,141	156
South Carolina	6,079	6,769	1,838,187	272	344,516	51	1,493,671	221

[a]Based on total number of cases ever under care.
Source: *Monthly Report of the F.E.R.A.*, August 1935, Tables B-2 and B-3, pp. 21-22.

APPENDIX B

METHOD AND SCOPE OF THE STUDY

METHOD AND SCOPE OF THE STUDY

A plantation was defined in this study as a tract owned or leased by one individual or corporation and operated under one management by five or more families, including that of the resident landlord. This is in conformity with the definition adopted by the Census of 1910 with the exception that the 1910 Inquiry included only those tracts operated by five or more share tenant families whereas this study included plantations with as few as four tenant families if the landlord also resided on the plantation. This study also included a few tracts operated by five or more renter families and a few operated by the landlord and four or more families of hired laborers. Although the rented and wage labor plantations were excluded from the 1910 Inquiry it was felt that a complete picture of the present situation demanded the inclusion of rented plantations as representative of the disintegration of large-scale operation, and of wage labor plantations as the survival of a method of operation which was universal soon after the Civil War and has gradually died out. The sample selected at random from representative areas averaged 85 percent operated by share tenants with an occasional wage hand family to look after the landlord's unshared operations. The other 15 percent were about equally divided between renter-operated and laborer-operated tracts.

The study focuses attention on the plantation as the unit, *i.e.*, the complete operations of the enterprise are studied and the shares allocated to the landlord and to the tenant are analyzed separately with some cross-analysis.

It was originally intended to sample 800 plantations—a number considered adequate to represent cotton plantations in the Southeast, if carefully chosen. Later exigencies of time and personnel dictated a 20 percent reduction in this number so that the final results are based on 646 schedules of plantations which contained 9,414 tenant and laborer families.

The sample was restricted to the Eastern Cotton Area in order to secure relative homogeneity in trends and costs of production and in tenant relations, these factors operating somewhat differently in the Western Cotton Area. The sample was, therefore, apportioned to the States of North Carolina, Georgia, Alabama, Louisiana, Mississippi, and Arkansas[1] on the following

[1] South Carolina was included in background statistics of this study but was excluded from the plantation sample owing to the fact that a large-scale farming study had been made the year before by the South Carolina Experiment Station. Unpublished data from this study were available.

basis (Figure 6, chapter I). The number of plantations in each of these States in 1910 was multiplied by the percentage which cotton production in 1930 was of production in 1910. The proportion which this weighted product formed of the sum of the weighted products determined the proportion of the sample allotted to the State. Table A shows the resultant distribution of the sample by States:

Table A—PLANTATIONS INCLUDED IN THE SAMPLE, BY STATES

State	Number of Plantations	Percent Distribution
Total	646	100.0
Alabama	154	23.8
Arkansas	89	13.8
Georgia	115	17.8
Louisiana	68	10.6
Mississippi	174	26.9
North Carolina	46	7.1

Within each State the sample was apportioned to areas according to the number of plantation counties in the area,[2] weighted by the percentage of tenancy in 1930. For example, having determined by the first step that 174 plantations were to be allocated to Mississippi, these 174 plantations were distributed among the Delta, Mississippi Bluffs, and Black Belt areas in proportion to the product of the number of counties in these

Table B—PLANTATIONS INCLUDED IN THE SAMPLE, BY AREAS

Area	Number of Plantations	Percent Distribution
Total	646	100.0
Atlantic Coast Plain	56	8.7
Uppér Piedmont	40	6.2
Black Belt (A)[a]	112	17.3
Black Belt (B)[b]	99	15.3
Upper Delta	50	20.7
Lower Delta	50	7.7
Muscle Shoals	22	3.4
Interior Plain	30	4.6
Mississippi Bluffs	47	7.3
Red River	28	4.3
Arkansas River	29	4.5

[a]Cropper and other share tenant majority.
[b]Renter majority.

areas multiplied by the percent of all farms operated by tenants. By this process the sample was distributed in 11 areas as shown by Table B.

Thus, the plantations selected for study were distributed proportionately by States and by the homogeneous areas of the Cotton Belt which extend across State lines.

[2]For designation of these areas, see chapter I and Figure 5.

Within each area, the sample was allocated so far as possible to representative counties chosen on the basis of percentage of tenancy, per capita income from agriculture, and the value of farm land per acre.

Within each area, counties were selected that represented the average of all counties in per capita 1930 gross farm income and in the percentage of tenancy in 1930 as shown by a frequency distribution of the counties (Tables C and D). Within each

Table C—PLANTATIONS INCLUDED IN SAMPLE, BY PER CAPITA GROSS FARM INCOME OF COUNTY IN 1930

Per Capita Gross Farm Income	Number of Plantations
Total	646
$100 to $125	79
125 to 150	96
150 to 175	177
175 to 200	104
200 to 250	90
250 to 300	57
300 and over	43

county, the sample was selected at random from one or more townships or other minor civil divisions, which were representative of the county according to the percent of tenancy and the value of farm land per acre.

Before beginning the enumeration of plantations in a given county, the supervisor of the study conferred with the County Demonstration Agent or the Rural Rehabilitation Supervisor relative to the townships chosen, and on the advice of the county

Table D—PLANTATIONS INCLUDED IN SAMPLE, BY PERCENT TENANCY IN COUNTY IN 1930

Percent Tenancy	Number of Plantations
Total	646
40 to 50	21
50 to 60	12
60 to 70	135
70 to 80	194
80 to 90	192
90 to 100	92

official made such changes as appeared necessary to obtain a representative sample.

To obtain schedules for the required number of plantations in a given county, each enumerator was assigned a township, or section of a township, and instructed to enumerate every plantation along a main road and its branch roads until he had obtained the number of plantations apportioned to the township.

In the enumeration of a given plantation, the procedure was to fill the schedule for the landlord first and then the schedules for the tenants. Interviews with tenants were not held in the presence of the landlord. On plantations that had between 4 and 10 resident tenant families, schedules were taken for all families where possible; on larger plantations a random sample, apportioned to the different tenure classes, was taken, approximately 15 tenant schedules being the maximum number taken on any one plantation. The following table (Table E) shows the representation of the various tenure classes among all tenant families on the 646 plantations and in the sample of tenant families enumerated.

Table E—TENANT FAMILIES ON SAMPLE PLANTATIONS AND TENANT FAMILIES ENUMERATED, BY TENURE STATUS

Tenure Status	Tenant Families on Sample Plantations		Tenant Families Enumerated	
	Number	Percent Distribution	Number	Percent Distribution
Total	9,414	100.0	5,171	100.0
Wage hands	1,581	16.8	867	16.8
Croppers	5,370	57.0	2,886	55.8
Other share tenants	1,394	14.8	716	13.8
Renters	863	9.2	653	12.6
Displaced tenants	206	2.2	49	1.0

To insure the greatest possible accuracy the enumerator checked the tenant's statement of cash after settling, subsistence advances, etc., with the operator's statement. Also, the operator's and tenant's statements of A.A.A. benefit payments received by them were checked against the records in the office of the county agent. All data pertaining to the relief history of the tenant were obtained from the records of the County Emergency Relief Administration. Values of crops sold were obtained insofar as possible from sales records. Values of home used products were assigned by securing the quantity consumed and multiplying by the local prevailing market price estimated by the Farm Demonstration Agent.

As soon as possible after a plantation was enumerated, the plantation and tenant schedules were edited by the supervisor in conference with the enumerator. Return visits were made to the plantation when necessary.

APPENDIX C

PLANTATIONS ENUMERATED, BY COUNTIES

PLANTATIONS ENUMERATED, BY COUNTIES

Total	*646*
Alabama	*154*
Bibb	18
Bullock	25
Calhoun	12
Elmore	18
Hale	29
Lauderdale	22
Lowndes	30
Arkansas	*89*
Crittenden	12
Miller	18
Phillips	20
Jefferson	14
Lincoln	7
Lonoke	6
Pulaski	2
Woodruff	10
Georgia	*115*
Carroll	12
Dodge	9
Hancock	13
Jenkins	18
Madison	12
McDuffie	20
Mitchell	18
Webster	13
Louisiana	*68*
Caddo	10
Concordia	19
Lincoln	10
Tensas	19
Webster	10
Mississippi	*174*
Adams	25
Carroll	17
Clay	14
Coahoma	22
Quitman	20
Warren	25
Washington	21
Yazoo	30
North Carolina	*46*
Anson	13
Cumberland	12
Edgecombe	17
Iredell	4

APPENDIX D

SCHEDULES AND INSTRUCTIONS

FERA FORM DRS 200

SURVEY OF PLANTATIONS

FEDERAL EMERGENCY RELIEF ADMINISTRATION
HARRY L. HOPKINS, ADMINISTRATOR

DIVISION OF RESEARCH,
STATISTICS AND FINANCE
CORRINGTON GILL, DIRECTOR

PLANTATION SCHEDULE

ENUMERATED BY________ EDITED BY 1.________ 2.________

1. OPERATOR'S NAME________ WHITE____ NEGRO____ TRACT NUMBER________

2. LOCATION OF PLANTATION: STATE________ COUNTY________ TOWNSHIP________

   ROAD________ NEAREST TOWN________ DISTANCE (MILES FROM)________

3. OPERATOR'S RES.: ON PLANTATION________ MILES AWAY________ ADDRESS________

4. NUMBER IN OPERATOR'S HOUSEHOLD (EXCLUDING SERVANTS OR BOARDERS)________ HOW MANY OTHER FARMS OWNED________

5. OTHER OCCUPATION OF THE OPERATOR________ INCOME FROM OTHER OCCUPATION________

6. DATE AND METHOD OF ACQUIRING LAND

(ENTER— INHERITED, BOUGHT, GIFT, FORECLOSED, MARRIAGE)

	HOW ACQUIRED	ACRES	DATE
OWNED			
TOTAL OWNED			
ADDITIONAL RENTED			
PLANTATION TOTAL			

7. LAND RENTED OUT________ LAND OPERATED________

8. VALUE OF FARM LAND________ VALUE OF OPERATOR'S RES.________ VALUE OF OTHER BLDGS.________

   VALUE OF ANIMALS________ VALUE OF MACHINERY________ TOTAL VALUE________

9. NUMBER OF VACANT HOUSES WHICH ARE HABITABLE OR COULD BE MADE HABITABLE FOR $50________

10. TYPE OF TENANT	ACRES OPERATED	NO. OF FAMILIES
WAGE HANDS		
CROPPERS		
SHARE TENANTS		
RENTERS (CASH AND STANDING)		
FAMILIES IN HOUSE WITHOUT CROP		
TOTAL		

11. TYPE OF LAND	ACRES
IN CROPS	
TILLABLE LAND IDLE	
PASTURE	
WOODS NOT PASTURED	
WASTE LAND	
TOTAL	

FARM YEAR BEGINNING 1934

CROP RECORD

12. CROPS	1 HARVESTED ACRES	2 QUANTITY PRODUCED	3 OPERATOR'S SALES			4 OPERATOR'S FAMILY USE		5 TOTAL TENANT SHARES
			PRICE	QUANTITY	VALUE	QUANTITY	VALUE	
1. COTTON, WAGES								
2. COTTONSEED, WAGES								
3. COTTON, CROPPERS								
4. COTTONSEED, CROPPERS								
5. COTTON, SHARE TENANTS								
6. COTTONSEED, SHARE TENANTS								
7. CORN, WAGES								
8. CORN FODDER, WAGES								
9. CORN, CROPPERS								
10. CORN FODDER, CROPPERS								
11. CORN, SHARE TENANTS								
12. CORN FODDER, SHARE TENANTS								
13. TOBACCO, WAGES								
14. TOBACCO, CROPPERS								
15. TOBACCO, SHARE TENANTS								
16. IRISH POTATOES								
17. SWEET POTATOES								
18. WHEAT								
19. OATS								
20. COWPEAS FOR SEED								
21. COWPEAS FOR HAY								
22. ALFALFA HAY								
23. PEANUTS								
24. SUGAR CANE								
25. SORGHUM								
26. SOY BEANS								
27. OATS, CLOVER, VETCH								
28.								
29.								
30.								
31.								
32. ORCHARD								
33. GARDEN								
34. TOTAL CROPS								
35. OTHER SOURCES OF INCOME, RENT REC'D.								
36. OTHER (SPECIFY)								

13. A.A.A. Benefits

		Landlord	Tenants
Cotton	(Rental		
	(Parity		
Tobacco	(Rental		
	(Parity		
Hog	(Rental		
	(Parity		
Other	(Rental		
	(Parity		
Total			

14. Livestock Products	Operator's Sales			Operator's Family Use	
	Price	Quan.	Val.	Quan.	Val.
Butter					
Milk					
Chickens					
Eggs					
Pork & Pork Products					
Beef					
Total					

15. Total Cash Income (12,13,14)________

17. Livestock (Plantation Owned)

Horses________

Mules________

Cows________

Calves________

Sheep or Goats________

Pigs________

Chickens________

Other________

Number above workstock kept in central barn or pastures________

Kept on Tenant Acres________

What items of feed were bought?

________ Why ________

________ ________

________ ________

________ ________

________ ________

16. Current Expenses

Items	Total Amount
Wage Hands	
Rations or Board	
Cotton Chopping	
Cotton Picking	
Miscellaneous Labor	
Cropper Occasional Labor	
Repairs, Dwelling	
Repairs, Barns, Fences, Implements	
Feed, grain	
Feed, Roughage	
Veterinary Fees, etc.	
Seeds, etc.	
Fertilizer, wages	
Fertilizer, soda T. & C.	
Insurance on property	
Taxes	
Rent	
Interest	
Ginning	
Total expenses	
Cash after settling, to tenant	
Tenant's share of expenses	

18. No. of families advanced subsistence________

Usual Mo. advance______No. of Mos.______

Total L'lord advances for subsistence________

Interest rate chg'd______Am't. Int.______

Commissary: Yes___No___Compulsory:______

Yes___No___

Profit, Am't.________Loss, Am't.______

19. Machinery: What machine jobs are chg'd

To tenant________Not chg'd________

Is machinery centrally controlled________

20. Number of Tenants' and Laborers' Families Living on Place and Size of Operations

	1935		1934		1933		1932		1931		1930	
	White	Negro	White	Negro	White	Negro	White	Negro	White	Negro	White	Negro
Total Families												
Wage Hand												
Croppers												
Share Tenants												
Renters												
Displaced												
Land in Crops												

21. Frequency of visits of non-resident landlord__________

Employment of overseer and other supervisory employees__________Salaries__________

22. Social Contribution of Plantation: Laborers' or Tenants' Doctor Bills__________

School Contributions______Church Contributions______Entertainments, etc.______

Relation to Law: Does l'lord pay fines_____How often_____Serve as parole sponsor_____

Is there a plantation burying ground______Funeral expense aid______

How often do tenants use plantation owned transportation for private use______

Transportation to town__________How often__________

23. Landlord Borrowing for Current Expense

Source	Time Used		Interest Paid	Security	Amount
	Borrowed	Repaid			
Bank					
Fertilizer Co.					
Merchant					
Government					
Total					

24. Landlord Debts

(excluding those for current crop)

Kind	December 31, 1933		December 31, 1934	
	Amount	Rate	Amount	Rate
Mortgage (Land				
Mortgage (Chattels				
Bank				
Merchant Note				
Open Account				
Government				
Other				

INSTRUCTIONS FOR ENUMERATING PLANTATION SCHEDULE

First, secure from operator a check list of all tenants who were on plantation in 1934. For those who have moved away, endeavor to fill tenant schedule from operator's information.

Item 1. Enter name of operator (owner or lessee). By operator is meant the individual (may be corporation or estate) who receives the rents and landlord crop shares. If the land is owned by a corporation or estate, enter name thereof above operator's name.

Item 2. Enter name of State, county, township, road, and nearest town, with distance therefrom.

Item 3. Operator's residence refers to 1934 abode. If on plantation, write *yes*, otherwise, *no*. Enter distance from plantation to operator's residence. If operator is not on plantation, defer enumeration until he can be interviewed.

Item 4. Enter number living in operator's house, exclusive of servants or lodgers, as of December 31, 1934. If operator owns other farms, enter number of such farms.

Item 5. Enter other occupation from which operator derived an income in 1934 and to which he devoted more than one-fourth of his time, and enter approximate proportion of total net annual income from such employment. Enter exact occupation such as "lawyer", "merchant", "mill owner", etc.

Item 6. If plantation was acquired by present operator in one unit, enter method of acquisition, size, and date of acquisition on one line. If acquired in several parcels, enter size and date of acquisition of each parcel.

Additional rented: Enter other connecting acreage operated as a unit with the owned acreage. Plantation total combines these two.

Item 7. *Land rented out:* Enter acres rented out to cash or standing rent tenants. *Land operated* is plantation total in item 6 minus land rented out.

Item 8. Enter values at conservative market value, not low assessed value or high speculative value. Enter only farm values, omitting gins, commissaries, and operator's residence if same is not on the plantation. If operator rents land, buildings, etc., enter separately in the margin the value of these items. Do not duplicate values by including the value of items rented in the value of items owned.

Item 9. Enter only houses judged to be habitable, or which could be made so with $50 or less repair.

Item 10. In subdividing land by tenure of farmer, enter only crop land, *i.e.*, all land planted to crops in 1934. Total here should equal crop land in item 11. Do not duplicate land planted to two crops. If operator cultivates an acreage with casual labor (not resident wage hands), enter the amount of the acreage in column "Acres Operated" opposite "Wage Hands" and enter "O" in column "Number of Families."

Item 11. Land "In Crops" should equal total of column "Acres Operated" under item 10. Under "Tillable Land Idle" include all cultivatable land not in woods and not in use for crops or pasture. Under "Pasture" include all pasture land, open and woods. Under "Waste Land" include swamp land, land occupied by buildings, ravines, lakes, etc. The total acres under item 11 should equal total under item 6.

Item 12. Enter under "Harvested Acres" (col. 1) and "Quantity Produced" (col. 2) all plantation crops for 1934 except those on acreage rented out for cash or standing rent.

For crops sold from plantation, enter separately sales of wage laborer crop, cropper crop, and share tenant crop. Renters' crop sales are entered only on the Tenant and Summary Schedules. Enter all sales of *shared* crops, whether sold by operator or tenants, under "Operator's Sales" (col. 3) and enter under "Total Tenant Shares" (col. 5) the value of the tenants' share of the total sales. Tenants' sales of *unshared* crops are entered on tenants' schedules. In case of a shared feed crop, where operator feeds his share and tenants sell their share, enter the total sales under both columns 3 and 5.

Crops fed to livestock should be entered only under columns 1 and 2; not under column 3 or 4. Enter under "Operator's Sales" (col. 3) any 1934 crops held for future sales; enter under "Operator's Family Use" (col. 4) any 1934 crops or goods processed from crops held for future family use.

Item 12, line 35. Enter rent received from land and buildings; do *not* enter rent received from animals, machinery, implements, gins, etc. Enter standing rent in terms of monetary value.

Item 12, line 36. Enter other income from the plantations and specify. Enter here interest received by the operator from advances for subsistence.

Item 13. Enter rental and parity payments to operator and tenants except renters. Check amounts reported by operator with records of county agent when operator is not certain about the amount of benefits he received. Remember that in many instances several checks were issued, and try to guard against omissions. Check operator's statement of benefits received by tenants with the total benefits report by tenants.

Item 14. Enter separately the value of operator's sales and home use of livestock products. If operator does not remember money return, endeavor to get accurate quantity and multiply by the prevailing 1934 price in town. Enter as sales unsold 1934 products held for sale; enter as home use unused 1934 products held for use, as for example, canned goods, etc.

Item 15. This item is the total of Operator's Sales (Item 12, column 3, line 34), Other Sources of Income, Rent Received (Item 12, line 35), Other Income (Item 12, line 36), A.A.A. Benefits (Item 13, total for both landlord and tenants), and Operator's Livestock Products Sales (Item 14).

Item 16. Enter total expenditures of operator for current 1934 operations. Do not include expenditures for capital improvements, such as, erections of new buildings, fences, drains, etc. "Wage Hands"—enter wages paid to *resident* (not casual) wage hands. "Rations or Board"—enter only for *resident wage hands.* "Cropper Occasional Labor"—enter wages paid to *resident* cropper and other share tenant families. "Interest"—enter interest on money borrowed to finance current 1934 operations; do not enter interest on long term debts. If there is an overseer, or if there are other supervisory employees, enter the amount of salaries paid. Include any other expenses not listed.

"Cash After Settling, To Tenants"—enter the difference between Total Tenant Shares (Item 12, col. 5, line 34) and Tenants' Share of Expenses plus subsistence advances. Inasmuch as cash due tenants may be credited against their back debts, this figure may not agree with total reported by operator. "Tenants' Share of Expenses"—enter the total amount of expenses paid by operator and charged to tenants.

Item 17. Enter the average number of horses and mules furnished during crop year by operator to tenants—wage hands, croppers, and other share tenants. Enter the number of other livestock as of December 31, 1934. Number of work stock kept in *central barns or pastures* means number kept up by landlord and assigned out to tenants. *Kept on tenant acres* means tenant kept and had practically full time use of animal. Explanation of why feed was bought should be concrete, such as, "Crop Failure."

Item 18. Enter the number of resident families—wage hands, croppers, other share tenants, and renters—whose living expenses are furnished by the operator (if merchant or other person instead of operator furnishes tenants, indicate this), either in a lump sum or by monthly advances. In cases where wage hands and/or renters are furnished, enter separately in margin. If interest charges are included in the total shown as advances for subsistence, indicate this. The amount of interest shown must also be shown in line 36 of Item 12.

Check whether or not the operator of the plantation has a commissary for supplying tenants, regardless of whether it is located on the plantation or in town, used exclusively or in connection with other plantations of same operator or in connection with plantations of other operators; and whether tenants must use commissary. Profit refers to profit on commissary. If plantations other than the one enumerated are served by the commissary, enter only the amount of profit derived from the plantation enumerated.

Item 19. This item refers to the use of plantation or hired machinery. Enter spring plowing and harrowing, ditching, terracing, etc. *Do not enter ginning.*

Item 20. Using the list secured from operator of the tenants resident on the plantation for each of the years 1930 through 1935, enter the number of tenants by status and color for each of these years. Enter the total land in crops for each year, 1930 through 1935.

Item 21. Enter the number of times (for example, daily, biweekly, weekly, monthly, etc.) during the crop season operator visits tenants for supervision. After "Employment of Overseer and Other Supervisory Employees", enter number of employees of the operator who supervise tenants for him and in parenthesis enter the number of persons in the families of these employees. After "Salaries", enter the total amount paid to those employees.

Item 22. Social Contributions of Plantations: Enter under each item only the money value *contributed* by operator in money, goods, or services.

Item 23. Operator's Debts: Two tables are given for operator's debts. The first (23) is borrowing for current farm expenses and the second (24) is for farm loans made for a longer period than a year or renewed from last year. If the operator borrowed the money for 1934 operations or "stands" for the tenants' accounts, enter his transactions under item 23—"Landlord's Borrowing for Current Expense." Do not include loans for personal use. The first column indicates the source of the loan. If more than one loan is secured from one source, enter below first loan and bracket the two. Do not duplicate. If fertilizer was bought from merchant on credit, enter under "Amount" on line with Merchant. Similarly, if money was borrowed from bank to buy fertilizer. If fertilizer was secured on credit from fertilizer company, enter after Fertilizer. If money was borrowed to operate a commissary serving several plantations jointly, endeavor to prorate the proportion of the loan assignable to this plantation.

Under *Time Used,* enter date of borrowing and date of repayment. In case of merchant credit which is taken up at intervals during the crop season, enter average time for which the account runs except where fertilizer is bought from merchant, in which case enter three-fourths of the time the account is open. Under *Interest Paid,* enter the actual dollars paid as interest charge (not 6 percent or 8 percent), except in merchant credit enter amount of difference between credit price and cash price.

Under *Security* indicate the manner in which loan was secured, such as, open account, endorsed note, note crop lien, chattel mortgage, etc.

The Operator's Debts in the second table should be mutually exclusive, *i.e.*, if bank holds a mortgage, enter only under mortgage. If bank holds a note, enter after bank. Similarly, if government agency owns a mortgage, enter after mortgage. In recording mortgage on land, indicate who holds the mortgage by letter (g) for government agency, (i) for individual, and (b) for bank. *Unpaid* borrowings for current 1934 operations should be shown in item 24 as a debt at end of 1934.

FERA FORM DRS 204

SURVEY OF PLANTATIONS

FEDERAL EMERGENCY RELIEF ADMINISTRATION
HARRY L. HOPKINS, ADMINISTRATOR

DIVISION OF RESEARCH,
STATISTICS AND FINANCE
CORRINGTON GILL, DIRECTOR

TENANT OR LABORER SCHEDULE

(FILL ONE FOR EACH FAMILY RESIDING OR WORKING REGULARLY ON FARM IN 1934)

1. NAME________ RACE______ PLANTATION NUMBER______

OWNER OPERATOR'S NAME________ WHAT KIN TO TENANT______

BIRTHPLACE: STATE________ COUNTY______

2. SHARE OF CROPS: COTTON____ COTTONSEED____ CORN____ TOBACCO____ OTHER____

3. HOUSE TENURE, 1934________ 4. 1934 STATUS______

5. AGE____ HOW LONG ON THIS FARM____ HOW LONG ABOVE STATUS____

6. NO. MOVES TO TOWN AND BACK____ 7. NO. FARMS LIVED ON SINCE FARMING___ NO. YRS. FARMED___

8. YEARS FARMED AS: OWNER___ RENTER___ SHARE TENANT___ CROPPER___ LABORER___

9. FINANCIAL RESULTS OF PAST 5 YEARS

(IF ACCURATE, USE ACTUAL FIGURES: OTHERWISE CHECK)

YEAR	LOST	EVEN	GAINED	DEBT AT END OF YEAR	
				TO LANDLORD	OTHER
1934					
1933					
1932					
1931					
1930					

10. IF SHARE TENANT OR CROPPER, CASH AFTER SETTLING: PROFIT______ LOSS______

11. ADVANCES FOR FOOD AND CLOTHING: AMT. MONTHLY____, MONTHS___ EXTRAS____ TOTAL____

12. LAND OPERATED: CROP________ OTHER________

13. LIVESTOCK OWNED BY TENANT, JANUARY 1935: HORSES AND MULES:____ COWS____

PIGS____ POULTRY____

14. RELIEF RECORD

YEAR	ENTER ACTUAL MONTHS		AMOUNT	
	RELIEF	REHABILITATION	RELIEF	REHABILITATION
1933				
1934				
1935				

15. NUMBER IN FAMILY______ SIXTEEN AND OVER: MALE______ FEMALE______

16. NUMBER ABLE TO HELP WITH FARM WORK______

17. CROP RECORD

	Acres Harvested	Quantity Produced	Crop Sold (if renter)			Home Use (All tenants and laborers)		A.A.A.
			Price	Am't.	Val.	Am't.	Val.	
Cotton								
Tobacco								
Peas								
Hay								
Sweet Potatoes								
Corn								
Garden Products								
Other								
Sub-total Crops								
Livestock Products								
Milk								
Butter								
Eggs								
Pork								
Chickens								
Sub-total livestock								
Grand Total								

18. Operating expenses not paid by operator:

Animal hire ______________ Veterinary fee ______________

Labor hired ______________ Rent ______________

Fertilizer ______________ Share of sub-tenants ______________

Seed ______________ Interest ______________

Feed ______________ Other ______________

Repair ______________ Total tenant expense ______________

Ginning ______________

19. Tenant income: ______________

INSTRUCTIONS FOR ENUMERATING TENANT OR LABORER SCHEDULE

Fill one schedule for each tenant and laborer family resident on this farm in 1934 and try to get operator's estimates on 1934 operations of tenants who have moved away since last year.

Consider as resident only tenants with crop agreement, laborers who are allowed a house and employed monthly, or other laborers or displaced tenants who are allowed a house but have no employment off the plantation furnishing the majority of their income.

Item 1. Race: Record only white or Negro, using letters W and N. Plantation number should agree with tract number on corresponding Plantation Schedule, and should preferably be entered at the time when the Plantation Schedule is numbered.

Item 2. Rental agreement, 1934: Enter 1/2 cotton, 1/3 corn, or other proportions according to what share of each crop the tenant receives.

Item 3. House Tenure, 1934: Refers to arrangement made with laborers without a regular agreement, or tenants without a crop, *i.e.*, a laborer housed on plantation or displaced tenant allowed use of house in return for casual jobs or small money rent, or old couple allowed rent free. Enter "rent free", "rent in return for odd jobs", etc.

Item 4. 1934 Status: Enter renter, share tenant, cropper, wage hand, Rehabilitation, resident no-contract. (See definitions in "Instructions to Enumerators"). To be determined by enumerator on basis of 2 and 3.

Item 5. Age: Enter as of birthday in 1934. *How long on this farm* refers to length of residence on the plantation as of 1934. Enter nearest whole number years only. If tenant has moved away and has come back without farming in the interim, enter the total years farmed on this plantation. If he has occupied another plantation in the interim, enter only last continuous occupancy of this plantation.

Item 6. Enter number of times tenant has moved to town and back since he started farming.

Item 7. Number of Farms Occupied: Enter the number of plantations occupied by tenant, plus the number of places he has farmed as an owner. *Number of Years Farmed:* Enter total years of tenant's farming experience since age 16.

Item 8. Enter total years in each status since 16 years of age, regardless of whether continuous or not.

Item 9. Get exact end results if tenant's memory seems trustworthy, *i.e.*, lost $57, or made $150. Gain is cash after settling, plus subsistence advanced, plus sales of unshared crops and livestock, plus A.A.A., plus wages, minus unshared expenses. In other words, the tenant may come out owing the landlord $50 but if he has been advanced $150 for subsistence his gain is $100. Debt includes chattel mortgages, balance unpaid on previous advances for subsistence, and bills due doctor, merchants, etc.

Item 10. Cash after settling is amount operator owed tenant after selling and dividing 1934 crops and deducting tenant's share of farm expenses and subsistence advances, or it may be the amount the tenant owes the operator as a result of the 1934 operation. The amount of this settlement applied to debts of previous years operation is not subtracted from this settlement but entered below.

Item 11. In recording advances enter average monthly advances, number of months the advance was allowed, extras allowed above average monthly advance, and total advances for the year.

Item 12. Enter total land operated by tenant, and crop land, *i.e.*, land planted in crops, including hay crops.

Item 13. Exclude plantation owned livestock and animals eaten or sold during the year. Record only those animals on hand at the end of 1934.

Item 14. Get general statement of this record from tenant and check later with County Relief Administration.

Item 15. Enter total persons in house on December 31, 1934. Enter the number of those 16 and over by sex, including head of household.

Item 16. Refers to number of persons 16-64 years of age of both sexes, including head of household, able to perform full day's labor.

Item 17. Crops sold need only be entered for renters unless other tenant does not sell crops through landlord. Enter amount sold and price, and multiply for value. Do not include sale of crops carried over from earlier years. In case of crops produced in 1934 and held for future sale, enter at average 1934 price. Enter all crops consumed by family and canned for future consumption. Omit crops fed to live stock.

Item 18. Enter only for renter or tenant not advanced these items by landlord.

Item 19. Tenant income is total of cash after settling, advances for subsistence, A.A.A., home use products, sales of unshared crops and wages minus unshared expenses.

DEFINITIONS OF TENURE STATUS

1. *Wage Hand*—An individual (with or without a family) who lives on the plantation and has a definite agreement with the operator to work for a more or less definite number of months at an agreed wage.
2. *Cropper*—A family which has a definite agreement with the operator whereby the family furnishes only labor (operator furnishes work stock and implements) in cultivating an agreed upon acreage and receives in return a specified share of the crop, usually one-half share or less.
3. *Other Share Tenant*—A family which has a definite agreement with a landlord whereby the family furnishes some or all of the work stock and implements in cultivating an agreed upon acreage and receives in return a share of the crop, usually more than one-half.
4. *Cash Renter*—A family which pays cash for the use of the land.
5. *Standing Renter*—A family which has a definite agreement with the operator whereby the family pays a specified amount of crop produce (for example, 4 bales of cotton, 800 pounds of tobacco, etc.) and which operates independently of the operator.
6. *Displaced Tenant*—A family in house without crop or definite work agreement and without regular outside employment which receives house rent free or in some instances gives a specified amount of labor in lieu of rent, but which has no definite agreement with a landlord to farm a crop or to work for wages for a specified period of time. Such a family may be allowed to cultivate a garden and crop patches but must depend upon casual labor for a cash income.

Note: A rehabilitation client was handled as a cash renter.

FERA Form DRS 201a 5491

SURVEY OF PLANTATIONS

FEDERAL EMERGENCY RELIEF ADMINISTRATION
Harry L. Hopkins, Administrator

DIVISION OF RESEARCH,
STATISTICS AND FINANCE
Corrington Gill, Director

SUMMARY SCHEDULE

Plantation Number________
Combined by________

A. Plantation Gross Shared Cash Income (15) ______

B. Plantation Shared Expenses (total) (16) ______

C. Plantation Net Shared Cash Income (A minus B) ______

D. Landlord, Home Use ______

E. Total Tenant Home Use (mark incomplete if all tenants are not enumerated) ______

F. Total Income ______

G. Tenant Share of A.A.A. (13) ______

H. Tenant Share of Crops (12 last col.) ______

I. Tenant Shared Cash Income ______

J. Tenant Crop and Livestock Sales (17) ______

K. Tenant Share of Expenses (16) ______

L. Net Tenant Shared Cash Income ______

M. Tenant Total Income (L plus J plus E) ______

N. Net Landlord Cash Income (C minus L) ______

O. Net Landlord Total Income (N plus D) ______

P. Value Less Residence (8) ______

Q. Six Percent of Value ______

R. Labor Income and Profit (O minus Q) ______

S. Crop Acres Operated ______

T. Per Acre Income $\frac{R}{S}$ ______

U. Landlord Household ______

V. Total Tenants' and Laborers' Households ______

W. Total Plantation Population ______

X. Per Capita Income $\frac{CC}{W}$ ______

Y. Tenant Unshared Expense ______

Z. Tenant Net Income (M minus Y) ______

AA. (Z plus O) ______

BB. Wages ______

CC. Income Plus Wages ______

DD. Gross Cash Income (A plus J) ______

EE. Plantation Kind Income (D plus E) ______

FF. Plantation Total Income (DD plus EE) ______

GG. Income per Dollar $\frac{FF}{P}$ ______

HH. Value Land per Acre ______

II. Percent Crop Land in Cotton ______

KK. Tenant Net Income per Family $\frac{Z}{10 - 2 \& 3}$ ______

Enumerators omit items DD to KK.

INSTRUCTIONS FOR SUMMARY SCHEDULE

Item A. Item 15 on Plantation Schedule.
Item B. Total of Item 16 on Plantation Schedule.
Item C. Subtract B from A. If B is greater than A, subtract A from B and place a ring around Item C.
Item D. Combine totals of "Operator's Family Use" Column, Item 12 and 14 on Plantation Schedule.
Item E. Total of "Home Use" column from Item 17 on Tenants' Schedules.
Item F. C plus D plus E.
Item G. Total of "Tenants" column from Item 13 on Plantation Schedule.
Item H. Total of "Total Tenant Shares" column from Item 12 on Plantation Schedule.
Item I. G plus H.
Item J. Total of "Crop Sold" column from Item 17 on Tenants' Schedules.
Item K. From Item 16 on Plantation Schedule.
Item L. I minus K.
Item M. L plus J plus E.
Item N. C minus L.
Item O. N plus D.
Item P. "Total Value" from Item 8, less value of operator's residence if residence is on plantation, on Plantation Schedule.
Item Q. Six percent of P.
Item R. O minus Q.
Item S. Total of first four lines from Item 10 on Plantation Schedule.
Item T. Divide R oy S.
Item U. Item 4 plus number of persons in household of overseer, etc., from Item 21, Plantation Schedule.
Item V. Total of Item 15 from Tenants' Schedules.
Item W. U plus V.
Item X. Divide CC by W.
Item Y. Total of Item 18 from Tenants' Schedules.
Item Z. M minus Y.
Item AA. Z plus O.
Item BB. Total wages (wages and rations to wage hands, wages to croppers, and wages to overseer and/or other supervisory employees) from Item 16 on Plantation Schedule.

Item CC. AA plus BB.
Item DD. A plus J.
Item EE. D plus E.
Item FF. DD plus EE.
Item GG. Divide FF by P.
Item HH. Divide "Value of Land" from Item 8 by "Total Owned" from Item 6, Plantation Schedule.
Item II. Divide "Cotton Acreage" from Item 12 by total number of acres, first three lines from Item 10, Plantation Schedule.
Item JJ. -
Item KK. Divide Z by total number of families, first three lines, Item 10, Plantation Schedule.

INDEX

INDEX